End User's Guide to Innovative Flexible Circuit Packaging

ELECTRICAL ENGINEERING AND ELECTRONICS

A Series of Reference Books and Textbooks

1. Rational Fault Analysis, *edited by Richard Saeks and S. R. Liberty*
2. Nonparametric Methods in Communications, *edited by P. Papantoni-Kazakos and Dimitri Kazakos*
3. Interactive Pattern Recognition, *Yi-tzuu Chien*
4. Solid-State Electronics, *Lawrence E. Murr*
5. Electronic, Magnetic, and Thermal Properties of Solid Materials, *Klaus Schröder*
6. Magnetic-Bubble Memory Technology, *Hsu Chang*
7. Transformer and Inductor Design Handbook, *Colonel Wm. T. McLyman*
8. Electromagnetics: Classical and Modern Theory and Applications, *Samuel Seely and Alexander D. Poularikas*
9. One-Dimensional Digital Signal Processing, *Chi-Tsong Chen*
10. Interconnected Dynamical Systems, *Raymond A. DeCarlo and Richard Saeks*
11. Modern Digital Control Systems, *Raymond G. Jacquot*
12. Hybrid Circuit Design and Manufacture, *Roydn D. Jones*
13. Magnetic Core Selection for Transformers and Inductors: A User's Guide to Practice and Specification, *Colonel Wm. T. McLyman*
14. Static and Rotating Electromagnetic Devices, *Richard H. Engelmann*
15. Energy-Efficient Electric Motors: Selection and Application, *John C. Andreas*
16. Electromagnetic Compossibility, *Heinz M. Schlicke*
17. Electronics: Models, Analysis, and Systems, *James G. Gottling*

18. Digital Filter Design Handbook, *Fred J. Taylor*

19. Multivariable Control: An Introduction, *P. K. Sinha*

20. Flexible Circuits: Design and Applications, *Steve Gurley, with contributions by Carl A. Edstrom, Jr., Ray D. Greenway, and William P. Kelly*

21. Circuit Interruption: Theory and Techniques, *Thomas E. Browne, Jr.*

22. Switch Mode Power Conversion: Basic Theory and Design, *K. Kit Sum*

23. Pattern Recognition: Applications to Large Data-Set Problems, *Sing-Tze Bow*

24. Custom-Specific Integrated Circuits: Design and Fabrication, *Stanley L. Hurst*

25. Digital Circuits: Logic and Design, *Ronald C. Emery*

26. Large-Scale Control Systems: Theories and Techniques, *Magdi S. Mahmoud, Mohamed F. Hassan, and Mohamed G. Darwish*

27. Microprocessor Software Project Management, *Eli T. Fathi and Cedric V. W. Armstrong (Sponsored by Ontario Centre for Microelectronics)*

28. Low Frequency Electromagnetic Design, *Michael P. Perry*

29. Multidimensional Systems: Techniques and Applications, *edited by Spyros G. Tzafestas*

30. AC Motors for High-Performance Applications: Analysis and Control, *Sakae Yamamura*

31. Ceramic Materials for Electronics: Processing, Properties, and Applications, *edited by Relva C. Buchanan*

32. Microcomputer Bus Structures and Bus Interface Design, *Arthur L. Dexter*

33. End User's Guide to Innovative Flexible Circuit Packaging, *Jay J. Miniet*

Additional Volumes in Preparation

Electrical Engineering-Electronics Software

1. Transformer and Inductor Design Software for the IBM PC, *Colonel Wm. T. McLyman*

2. Transformer and Inductor Design Software for the Macintosh, *Colonel Wm. T. McLyman*

End User's Guide to Innovative Flexible Circuit Packaging

Jay J. Miniet
Motorola, Inc.
Plantation, Florida

MARCEL DEKKER, INC.   NEW YORK · BASEL

Library of Congress Cataloging in Publication Data

Miniet, Jay J., [date]
End user's guide to innovative flexible circuit packaging.

(Electrical engineering and electronics ; 33)
Bibliography: p.
Includes index.
1. Flexible printed circuits. I. Title. II. Series.
TK7868.P7M54 1986 621.381'74 86-13373
ISBN 0-8247-7584-8

MARCEL DEKKER, INC.
270 Madison Avenue, New York, New York 10016

Current printing (last digit):
10 9 8 7 6 5 4 3 2 1

Printed in the United States of America

This book is dedicated to the Motorola Secure Communications Management who stood by my innovative ideas from conception through production.

PREFACE

In the last decade, the electronic industry has witnessed the use of flexible circuits expand from point-to-point wiring replacement to the current broad spectrum of uses. This growth has been apparent in many electronic market segments: the military, computer and consumer products, and communication equipment. In these markets, flexible circuits have proven their value in use by increasing reliability, reducing manufacturing cost, saving weight and space, reducing assembly cost, and providing greater design freedom in numerous applications.

Although flexible circuits have distinct advantages over conventional round wire harnesses and printed wiring boards in many applications, flexible circuits now represent less than ten percent of all the printed wiring produced. Many potential end uses for flexible circuits are waiting for innovative design engineers to apply the flexible circuit technology. Industry inquiries from electronic design engineers verify the need for broader dissemination of knowledge about the design advantages and proper application of flexible printed circuits.

This book addresses these engineering needs. The main purpose is to guide design engineers through the different steps required to design flexible circuit packaging.

There are chapters that review not only the comparison between flexible circuits, printed circuit boards, and hybrids, but also the design considerations to keep in mind to design flexible circuit packaging with surface mounted devices and leaded components.

It is the hope of the author that you will find in this book not only the answers to your questions, but also the motivation to further the development of new design techniques and better flexible circuit packaging. The author is employed by Motorola Communication Division and is familiar with portable, two-way communication equipment. You will therefore see references to portable radios and the "Flex-O-Pak," a flexible circuit packaging technique developed by the author.

Special thanks to Dana Allensworth, Larry Leicht, and Glenn Gilstrap for their technical support and especially to my daughter, Elizabeth, for her indispensable support in the writing of this book.

Jay J. Miniet

CONTENTS

1
DESIGN CRITERIA

1.1 Circuit Substrate Alternatives

Until recently, the electronic industry relied mostly on rigid printed wiring boards (PWBs) and ceramic hybrids for its packaging. Although these two methods have proven invaluable to the industry, they are 180° apart from each other.

Rigid boards

Some of the most popular types of printed wiring boards are double-sided boards with plated-through-vias, and multilayer boards. These PWBs are almost exclusively fabricated from FR-4, a substrate which is made up of fiberglas and epoxy resin, but some CEM and CRM grades are making inroads because of price and ease of drilling and punching. FR-4 and G-10 remain today's most popular substrate laminates for computer and telecommunications, as well as for military and aerospace equipment.

FR-4 has the combination of properties that best meets the needs of the PWB fabricator and user. For example, FR-4 and G-10 remain flat, are dimensionally stable, have resistance to heat and processing chemicals, and machine well. There are laminates that are superior in any one of the above characteristics, but they do not offer the combined benefits of the FR-4 laminate (see Table 1).

The PWB industry's efforts have been aimed at improving the laminate's heat resistance through better cross-linking of the resin system, thereby raising the glass transition temperature (Tg) (Fig. 1). The Tg has become a measure of how well a laminate resin system resists softening from heat. When the Tg temperature is reached, the resin changes from its glassy state and causes changes in the laminate's properties. The Tg is not a measure of the resin's melting point, but rather a point at which molecular bonds begin to weaken enough to cause a change in physical properties (dimensional stability, flexible strength, etc.). Its value is determined by the intersection of the two slopes of the temperature/property-change curve. The FR-4 epoxy systems have a Tg of 115–125°C; polyimide resin systems, 260–300°C.

Although the Tg value is measure of the estability of a material under heating, it is not meant for comparison purposes once the Tg is exceeded. For example, materials with a higher Tg are desirable because they maintain their stability over a wider temperature range up to the Tg. However, once the Tg is exceeded, the properties of a high Tg resin could change much more rapidly than a material with a low Tg. At a higher temperature, the lower Tg material may even exhibit superior properties (see Fig. 2).

Epoxy laminates provide the best dimensional stability, especially constructions employing woven glass reinforcement and reduced resin content. Epoxy has a low shrink-

Table 1 Laminate Designations and Materials

	Resin			Reinforcement				
Grade	Epoxy	Polyester	Phenolic	Cotton paper	Woven glass	Mat glass	Glass veil	Flame retardant
XXXPC			•	•				
FR-2			•	•				•
FR-3	•			•				•
FR-4	•				•			•
FR-5	•				•			•
FR-6		•				•		•
G-10	•				•			
CEM-1	•			•	•			•
CEM-2	•			•	•			
CEM-3	•				•	•		•
CEM-4	•				•	•		
CRM-5		•			•	•		•
CRM-6		•			•	•		
CRM-7		•				•	•	•
CRM-8		•				•	•	

Courtesy of Electronic Packaging and Production, June, 1984.

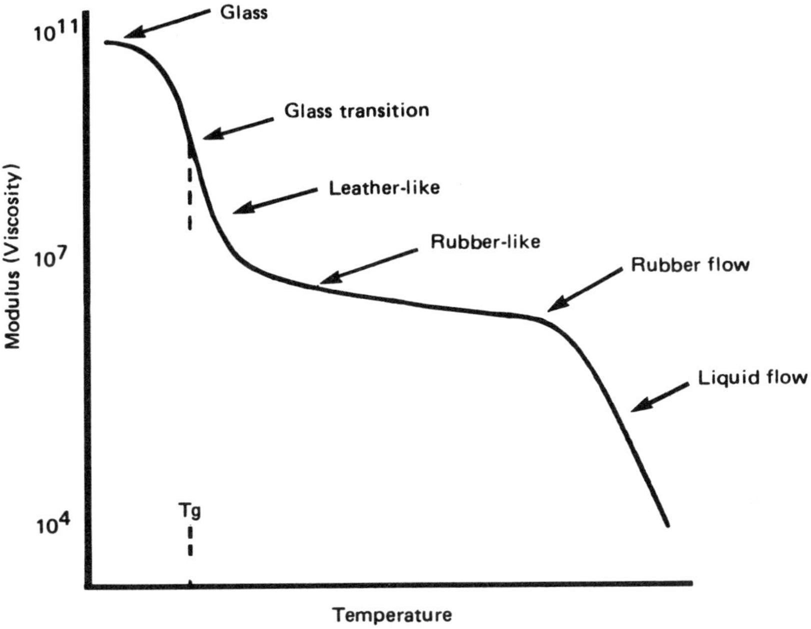

Fig. 1 Glass transition temperature.

age factor, and the continuous glass fibers offer resistance to tensile stress.

Polyimide and BT epoxy resin systems are the choice when thermal stability is a major requirement. These resins have a higher Tg, lower Z-axis expansion, and maintain peel strength at elevated temperature (Fig. 3). Polyimide/glass laminate construction, as well as cor-lam, are now specified for many military programs.

The PWB industry, especially the military segment, has been addressing the thermal coefficient of expansion (TCE) mismatch problem existing between SMDs (primarily LCCC) and the supporting substrate by using differ-

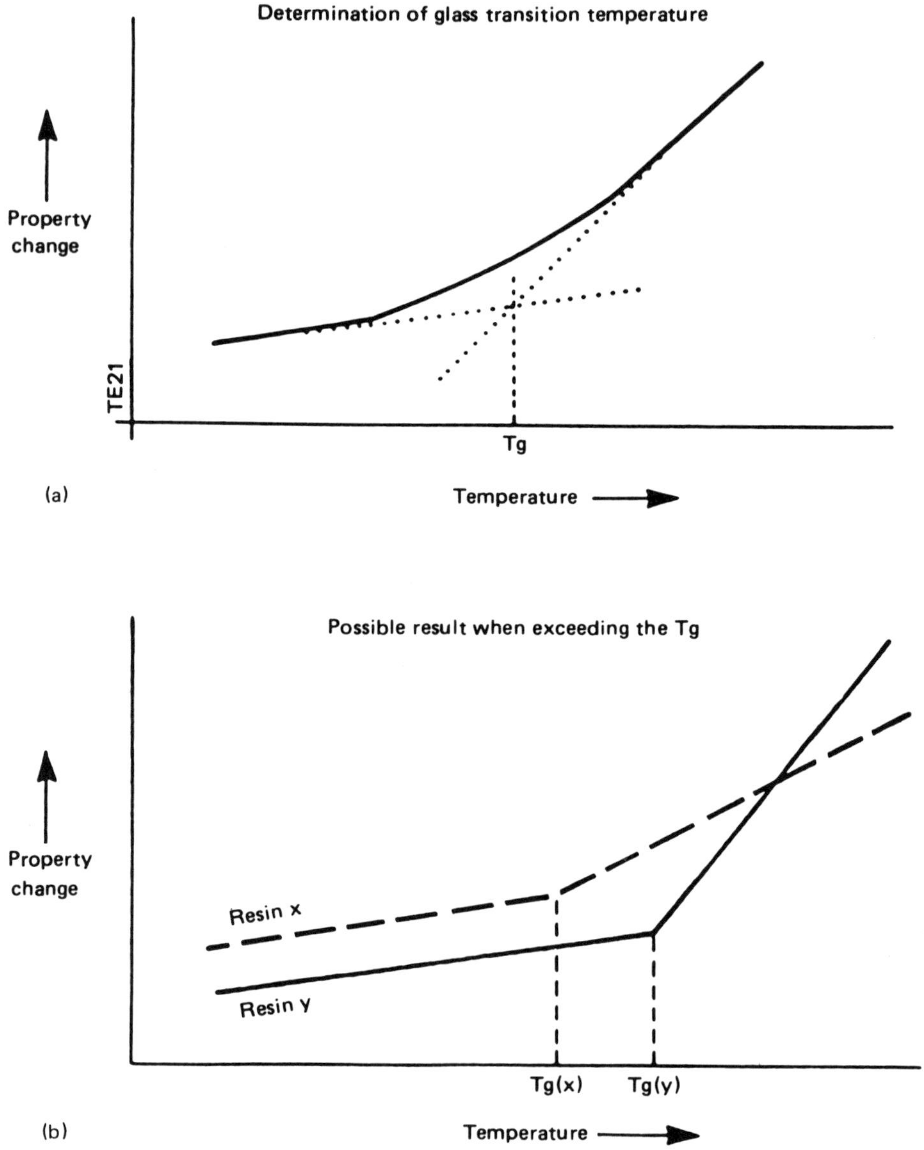

Fig. 2 The laminate's glass transition temperature (T_y) is the point where properties exhibit significant change (a). The higher the T_g the more dimensionally stable the laminate—except when exceeded, then the T_g becomes meaningless (b). Courtesy of *Electronic Packaging and Production*, June, 1984.

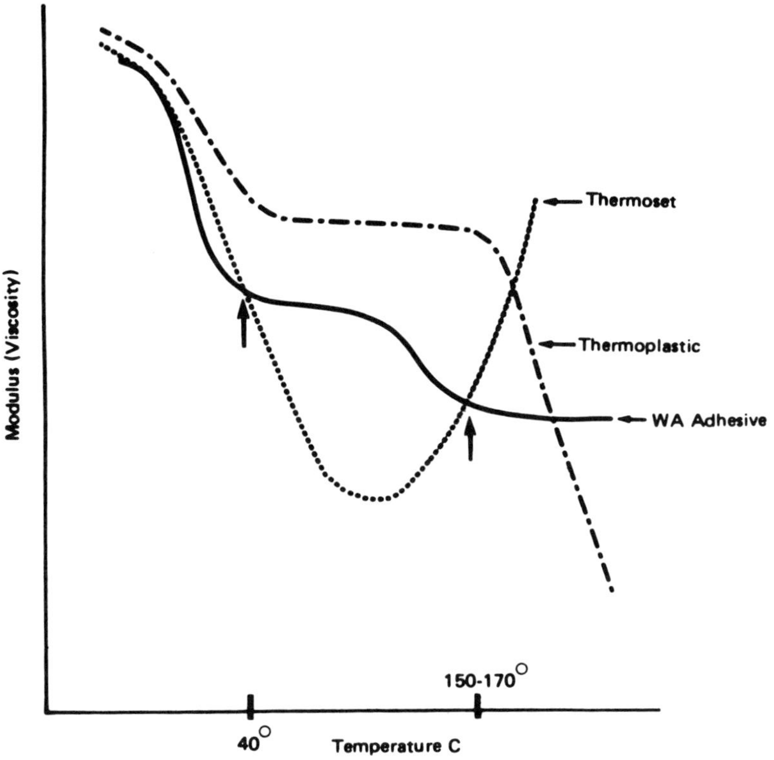

Fig. 3 Flow characteristics showing thermal stability.

ent reinforcement materials such as Kevlar with a TCE of (−2 to −4 ppm/C°) and quartz fibers (0.54 ppm/C°) instead of glass (4–5 ppm/C°).

All of these efforts have succeeded in reducing the TCE mismatch of the finished board so that it is closer to ceramic SMDs (5–7 ppm/C°). However, there is concern over the economics involved, since a processed board can cost fifteen times that of an FR-4 board.

As an alternate to Kevlar and quartz reinforcements, TCE control for multilayer boards is being addressed by the use of internal power plane materials having a low TCE, such as copper-clad Invar (5 ppm/C°). Thin epoxy/glass laminates (17 mils total thickness) clad on both sides with copper-Invar-copper have been successfully used as a core material for multilayer board construction. Polymide/glass with copper-clad Invar has proven to be equally successful.

A major factor involving the determination of dimensional and thermal stability is the lack of a standard, accurate, repeatable test capable of being reproduced in different test labs.

Hybrids

A single hybrid can serve as a small, lightweight substitute for PWB, or a collection of PWBs, with the added benefit of a considerable increase in reliability.

The major drawback is that tooling or front-end costs are required, and these costs are appreciable. Although hybrids have been widely used in aerospace and military applications, their commercial uses have been confined primarily to products produced in large volumes.

Among substrate materials Alumina and Beryllia ceramics still predominate, though competitive materials such as porcelain-coated steel may challenge this leadership as the size of the hybrid module market continues to grow. Polymer-based synthetics are also beginning to make an impact in low-cost commercial and consumer applications.

Alumina is actually a ceramic of aluminum oxide crystals, usually 95 or 96%, and calcium and magnesium oxides. This formulation offers excellent dieletric properties, and provides a smooth, partly-granular surface with a low coefficient of expansion, good tensile strength, and a hardness approaching that of sapphires and diamonds. The small metallic content, moreover, ensures good thermal conductivity and excellent heat dissipation.

Beryllium oxide is another ceramic for high-power applications. A ceramic of 99.5% beryllia is somewhat lighter in weight than alumina but costs more. Its prime advantage is somewhat better heat conductivity. A ceramic of 96% alumina, for example, has a thermal conductivity rating of 0.05 cal/s-°C-cm (at 25°C) versus 0.62 cal/s-°C-cm for beryllia. In other words, beryllia will conduct as much as twelve times the heat, measured in calories, as alumina.

However, alumina, beryllia, and other ceramics have their disadvantages. In spite of their high tensile strength, ceramics are brittle. They can therefore be shattered by impact, or cracked under repeated vibration or mechanical stress.

Hybrid fabrication methods have reached several levels of sophistication, some of which require only small front-end investments and thus are more suitable for both high- and low-volume applications. Inexpensive hybrids do not offer the same degree of space savings as the more costly varieties. This drawback can be offset by shorter development time and cost savings arising from the fact that they demand less careful handling and testing.

Hybrids are used for their improved reliability over printed wiring boards in such applications as cardiac pacemakers, telecommunications, automotive braking systems, and computer circuits. Hybrid reliability is generally said to be ten times greater than an equivalent discrete circuit board. Hybrids use some of the same mechanisms that increase reliability in LSI chips, and they also can be tuned or adjusted much like discrete circuit boards.

Reliability of discrete components mounted on circuit boards is affected by primarily by two things. First, solder connections on the board can fail from shock and vibration. Second, semiconductor components are lot sample tested by their manufacturer, and some small fraction of bad components is sent to users. Thus, the user must test each component to ensure reliability. Hybrid interconnection methods reduce these problems. To eliminate solder connections, some monolithic hybrids have components formed within the substrate as in an LSI chip. But more often, components are adhesive bonded to the substrate. These conductive adhesives are also more reliable than soldered connections. Solder connections are used on some hybrids, but far fewer are required than with discrete components on circuit boards. Most hybrids are encapsulated in epoxy, which further reduces the effects of mechanical vibration. In addition, encapsulating the hybrid at curing temperatures of approximately 150°C also provides a thermal test that weeds out weak semiconductors.

Thick-film hybrids

Hybrid circuits are of three types, each suited to specific uses. In *thick-film hybrids*, conductive and resistive pastes are deposited on a substrate and heated. The heating hardens the pastes and fuses them to the substrate. Once hardened, the resistive pastes become thick-film resistors, so named because the thickness is about 1 mil. Capacitors, discrete semiconductors, and ICs are then added to the circuit (see Fig. 4).

Thin-film hybrids

Thin-film hybrids are made by placing the substrate in a vacuum chamber and depositing films of conductive and resistive material on the entire substrate. The material is photoetched in selected patterns to form conductive interconnections and thin-film resistors. These resistors are approximately 0.1 mil thick. Uncapsulated components are then added either by soldering or adhesive bonding (see Fig. 5).

Monolithic hybrids

The third type are *monolithic hybrids*. The circuit is built on a multilayer semiconductor substrate. Semiconductor elements are actually formed within the substrate in a process much like that used to build a monolithic IC. Additional components can be soldered or adhesive bonded to the substrate, but usually components are fabricated within the substrate (see Fig. 6).

Of the three types, thick-film hybrids are probably the most widely used. Thick-film circuits can be separated into two distinct types on the basis of whether or not added components are prepackaged (encapsulated).

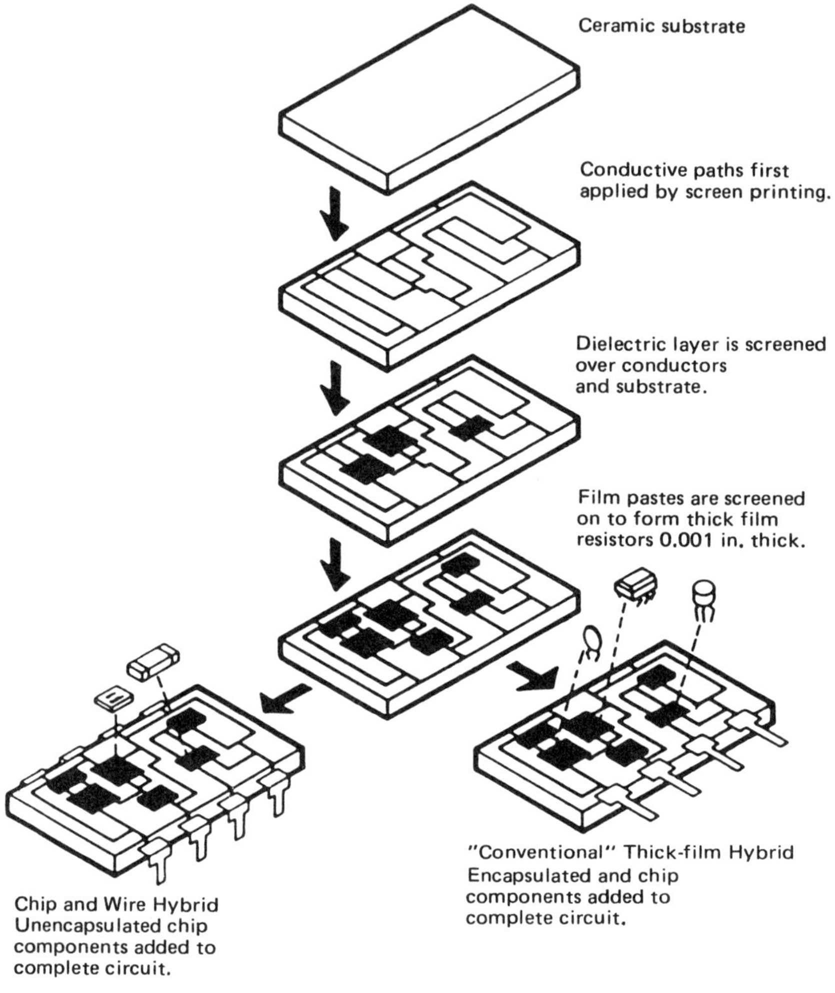

Fig. 4 Thick-film hybrids. Courtesy of *Machine Design*, June, 1978.

Thick-film hybrids which have added components (such as capacitors and ICs) already inside their own plastic packages are the least costly to develop. Added components are soldered conventionally to the substrate conductors. These circuits are frequently used

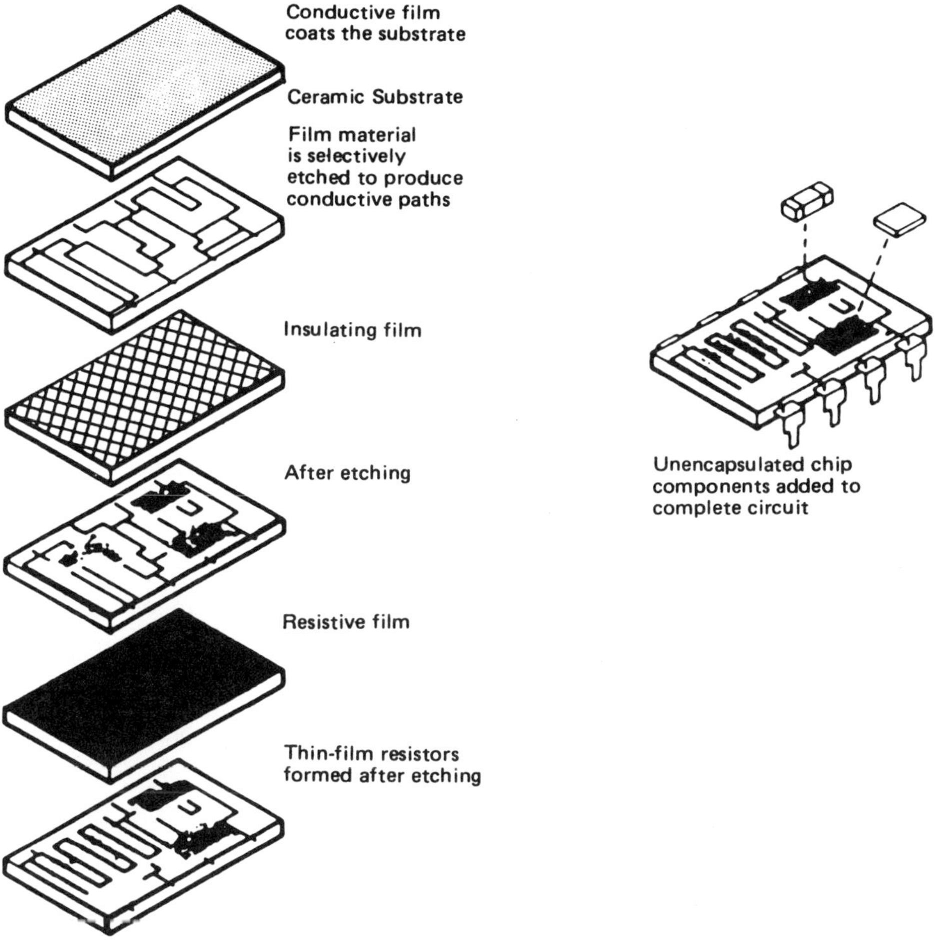

Fig. 5 Thin-film hybrids. Courtesy of *Machine Design*, June 1978.

for applications where low tooling costs are crucial.

On the other hand, a chip and wire thick-film hybrid uses unencapsulated components that usually are adhesively bonded to the substrate. They are electrically connected to the substrate conductors by minute welds

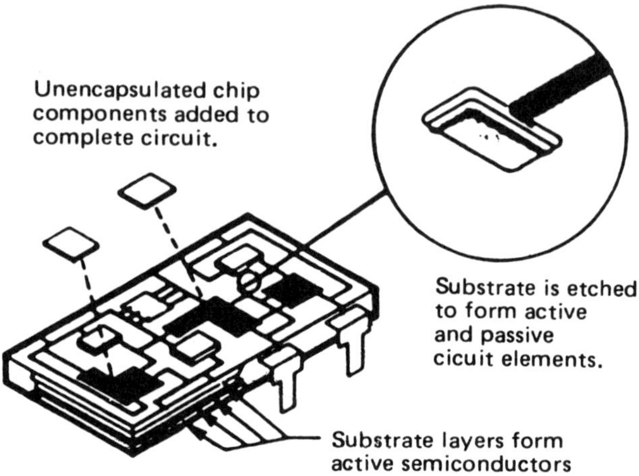

Fig. 6 Monolithic hybrids. Courtesy of *Machine Design*, June, 1978.

called wire bonds. The chips and other attachments used in the chip and wire process are less expensive than prepackaged components, but all handling, soldering, and wire bonding on the tiny devices usually must be done manually and with the aid of a microscope, thus, handling costs can be quite high.

These costs also affect the testing of complete hybrids. Unpackaged chips must be wire bonded to the conductors before they can be tested at normal operating voltages and frequencies. Chips that fail during testing must be meticulously unwelded, removed, and replaced in a time-consuming process. Prepackaged components can be easily pretested before being placed on the substrate. If a prepackaged component fails on the hybrid, it can be unsoldered and quickly replaced.

Perhaps the most significant advantage

of prepackaged components is the wide range of circuit technologies that can be used for attachment. Dissimilar components, such as PMOS, CMOS, and optoelectronic devices, often cannot be wire bonded on the same substrate because each requires a different temperature and process for adhesive bonding, and temperatures for one technology can be incompatible with those of another. Thus, hybrids using wire bonded chips are often limited to only one chip technology.

Flexible circuits

More recently, a third packaging method was introduced—flexible circuit packaging—to the electronic industry. Flexible circuit packaging, with its limitless adaptability to applications, continues to draw significant attention because of its many attractive features (see Table 2). Flexible circuitry offers packaging engi-

Table 2 Flexible Circuits—What Are They?

Features	Flexible circuit/ Cable types	Typical construction
Compact interconnection and packaging of electronic systems	Single-sided	Dielectric film[a]
Lightweight	Double-sided	Adhesive[a]
Flexible	Multilayer: flexible; rigid/flexible	Metal conductors[b]
Reliable		Adhesive[b]
Reduced assembly costs		Dielectric film[b]

[a]Cover layer
[b]Metal clad flexible laminate

neers greater conceptual freedom in their designs, increased reliability, weight and space savings, and reduced manufacturing and assembly costs. Flexible circuits are made in three basic types and are modified to satisfy a variety of unique packaging requirements.

Single-sided flexible circuits

The single-sided flexible circuit (Figs. 7 and 8) represent the simplest, and easiest to manufacture, member of the family. It is commonly constructed of a single layer of copper foil laminated to a base layer of insulating dielectric material. Another layer of insulating dielectric material, called a cover coat, may be laminated to the copper foil, encapsulating the conductors.

Double-sided flexible circuits

The double-sided flexible circuit (Fig. 9) is made of a base layer of dielectric material with copper foil laminated to both surfaces with adhesive. A cover coat layer of insulating material is laminated to both sides, encapsulating the conductors after drilling, plating and etching (Fig. 10).

Multilayer flexible circuits

Multilayer flexible circuits are perhaps the most effective mechanism for producing highly complex and densely packaged circuitry. Multilayer flexible circuits consist of three or more layers of copper foil interspaced with dielectric material, and laminated together with an acrylic adhesive or a modified epoxy adhesive (Fig. 11a). Interconnection between layers is accomplished with plated-through-

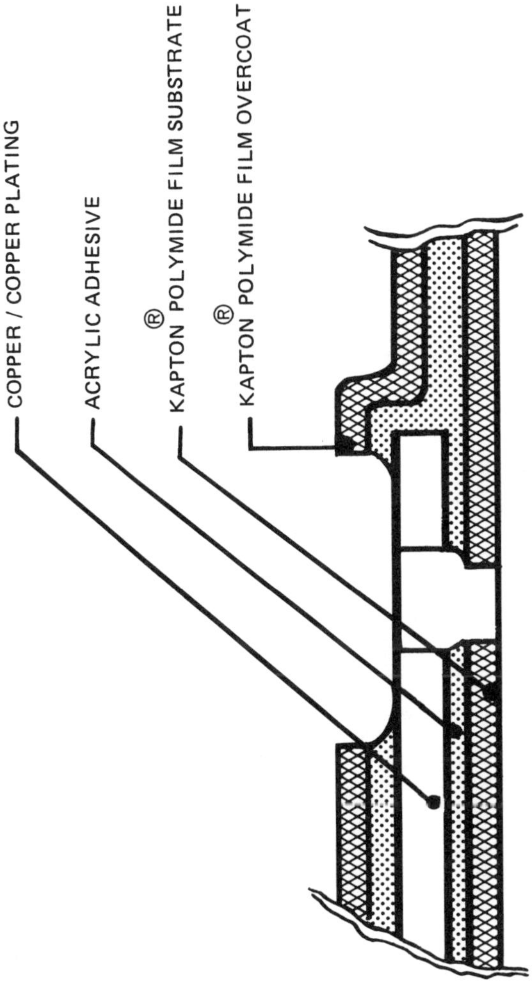

Fig. 7 Single-sided flex circuit.

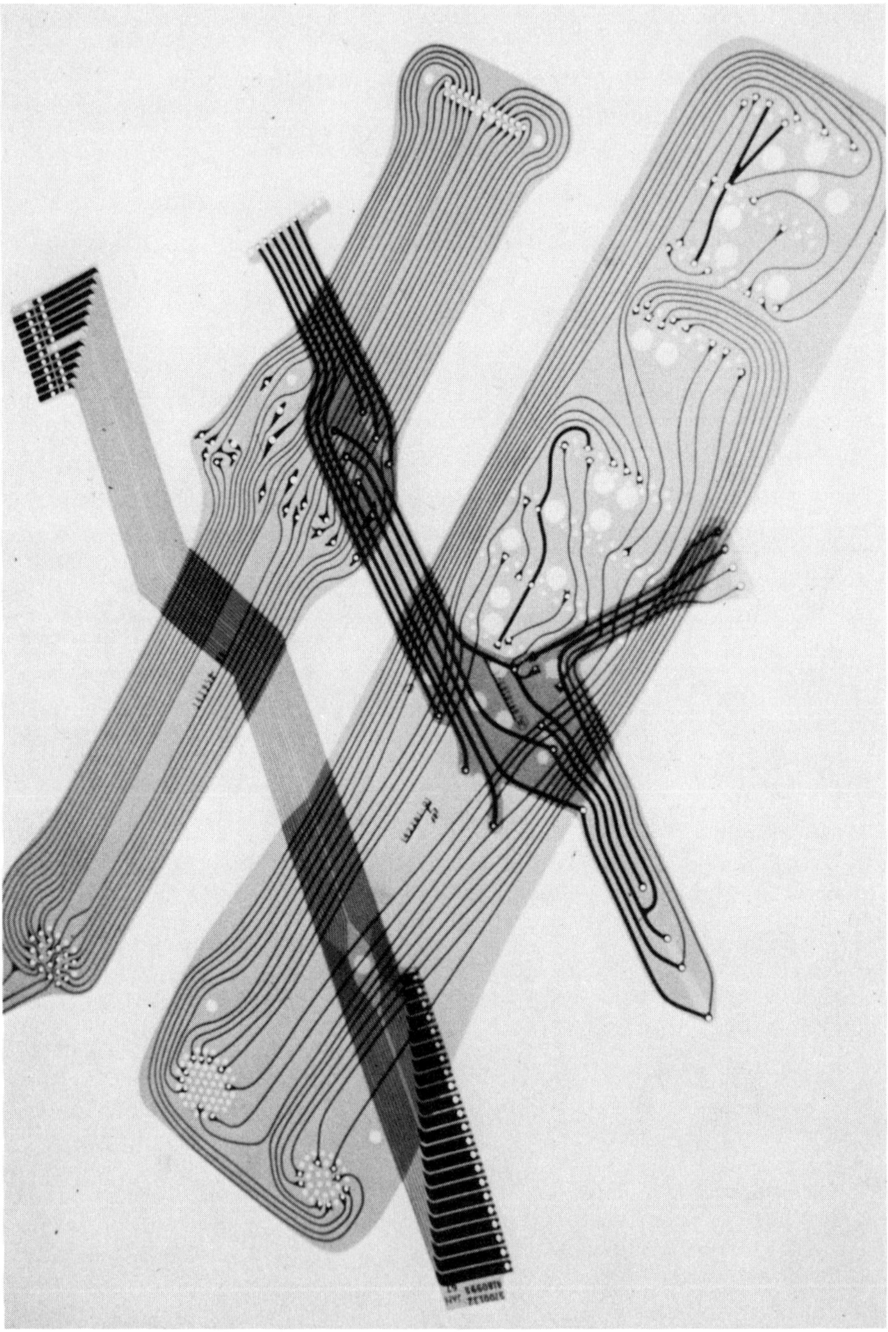

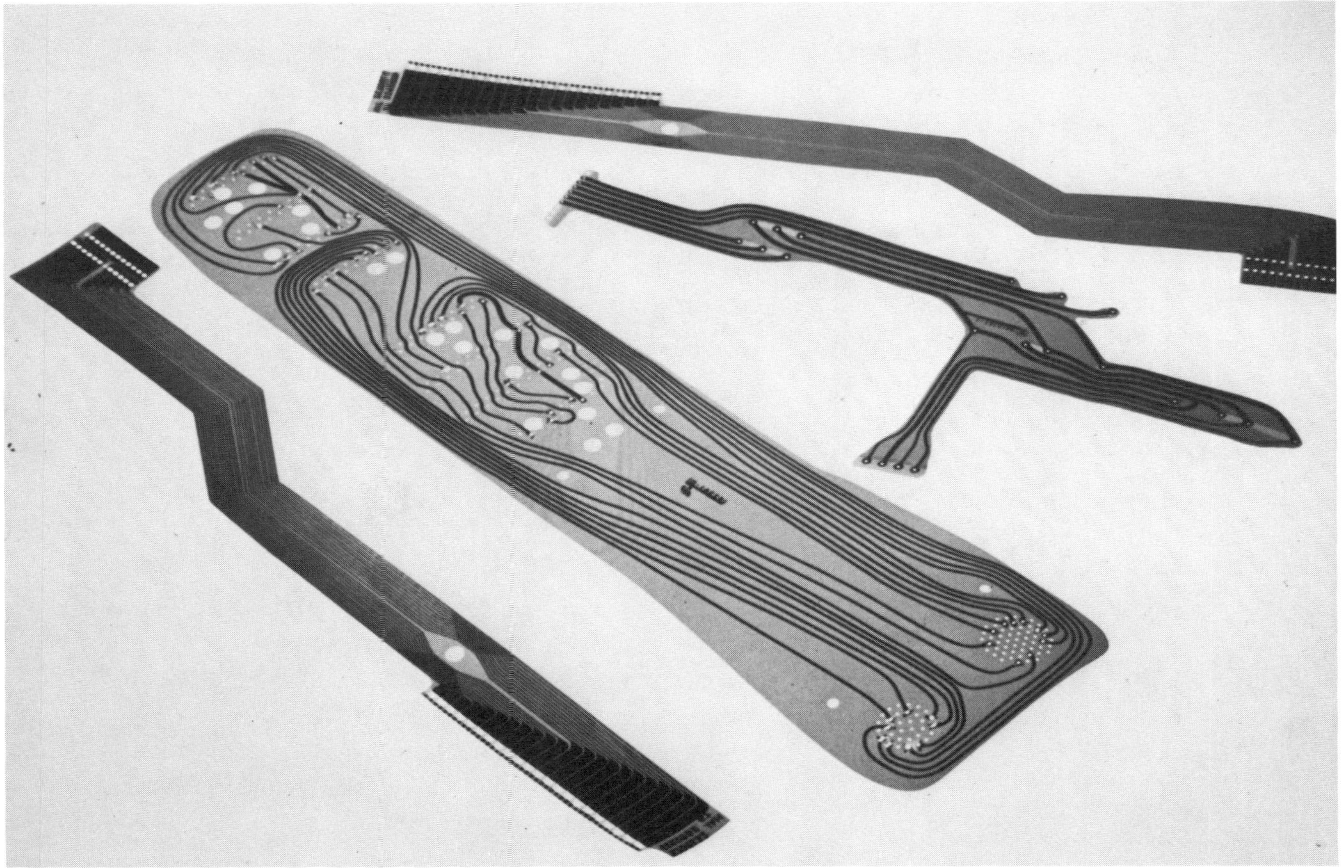

Fig. 8 Assorted single-sided flexible circuits. Courtesy of DuPont.

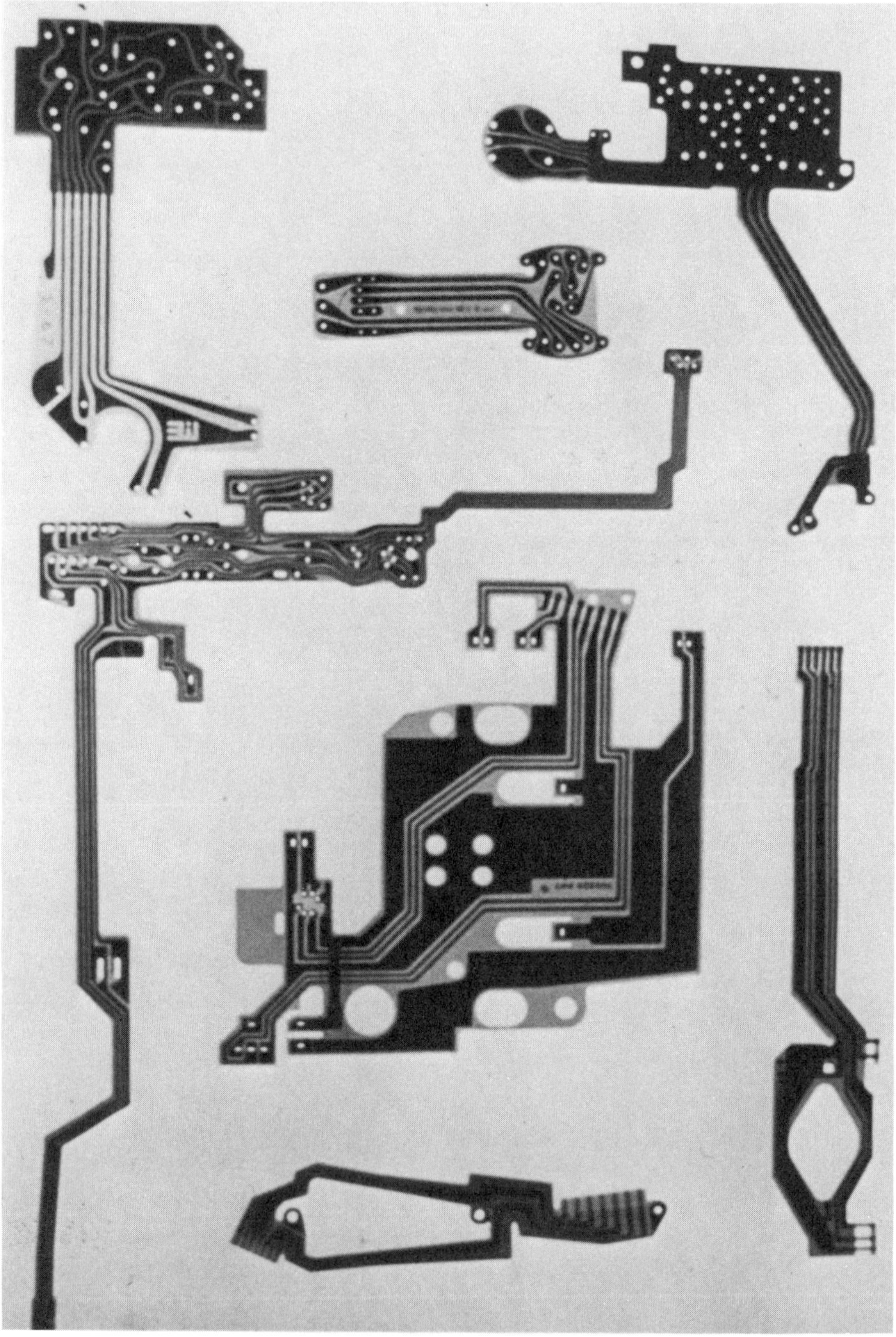

Fig. 8 (continued)

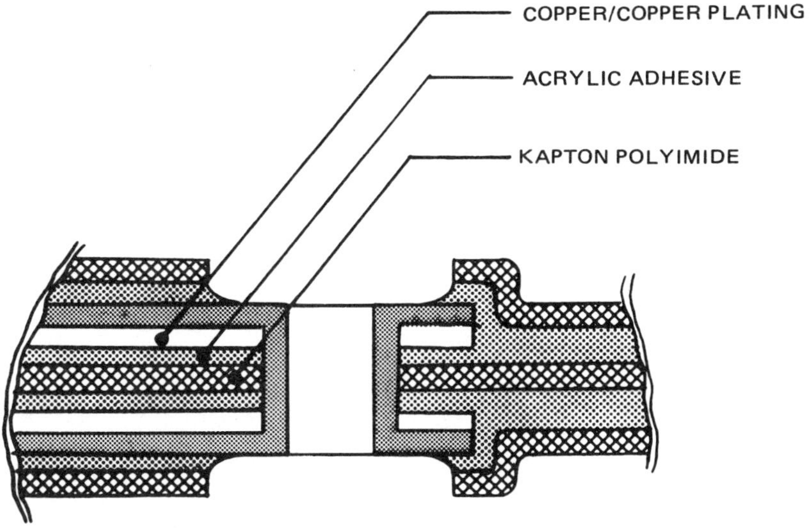

Fig. 9 Double-sided flexible circuit.

holes similar to double-sided flexible circuits (Fig. 11b).

Interconnection between layers is typically done with drilled, plated-through-holes. A new interconnect technology is being developed by DuPont Photosystems and Electronic Products Department, which forms the required layer-to-layer interconnects by photographic and chemical techniques, rather than by drilling.* This development offers the industry substantial technical and economic advantages in producing high-density microelectronic systems. In much the same way as architects design higher and higher buildings when real estate is limited, this new

*Reference: U.S. Patents 4,517,050 and 4,472,238.

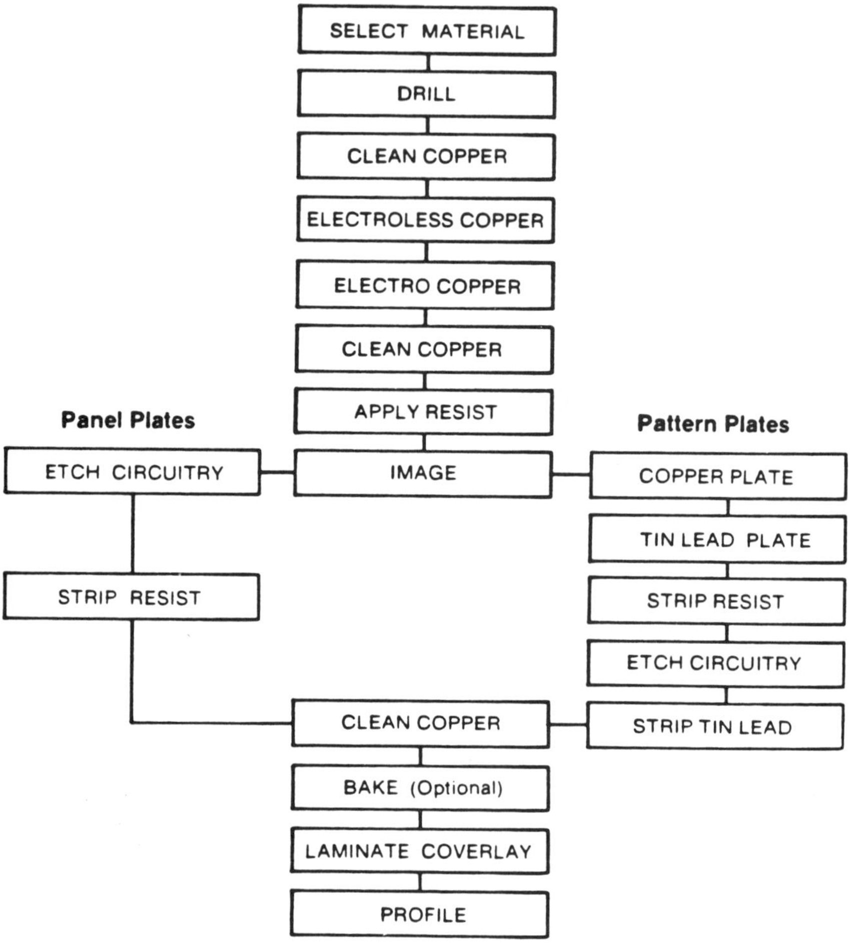

Fig. 10 Processing PYRALUX WA/K flexible composite double-sided and multilayer circuits with plated-through-holes panel and pattern plates.

system allows electronics engineers to crowd one layer of circuitry on top of another and another and another. Traditionally, these layers of circuitry have been connected by vias, or vertical holes, containing a conducting material. As packages are becoming more dense-

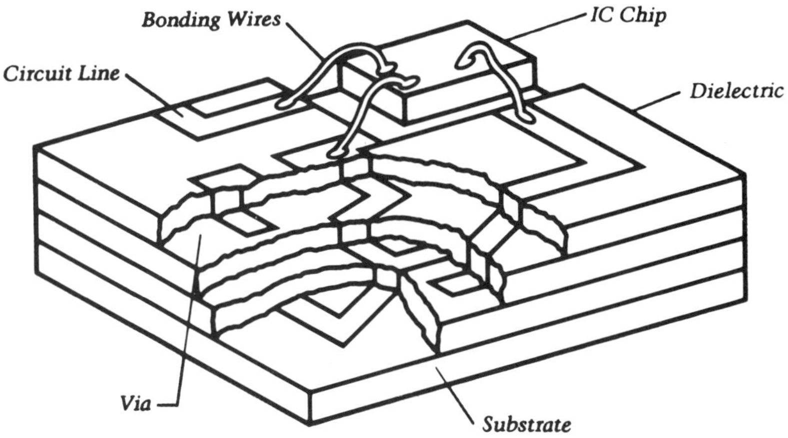

(a)

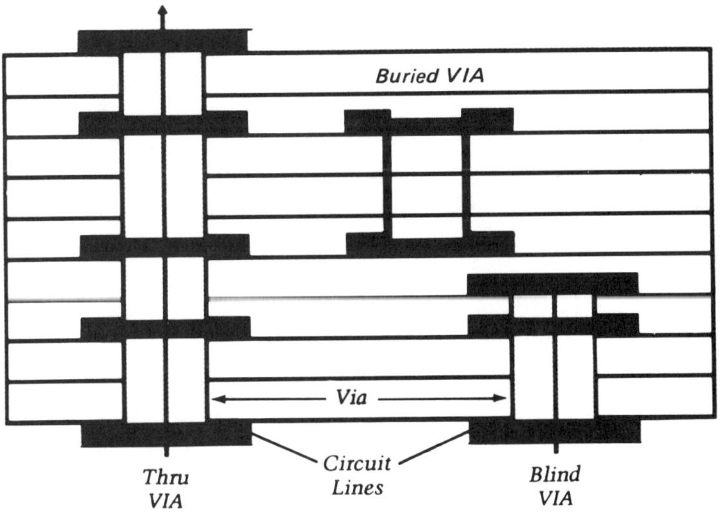

(b)

Fig. 11 Multilayer flexible circuit: (a) Multilayer structure; (b) Multilayer interconnect vias.

ly crowded with electronic components, the need has arisen to reduce the diameter of vias below the size attainable through drilling or punching. The development of advanced microelectronic packaging systems, such as that shown schematically in Fig. 11, involve new processes that will enable you to design sophisticated, high-precision multilayer parts.

Variations of the commonly used multilayer flexible circuit include designs with all flexible layers laminated together, or with a hard board backer area with one or more sides of FR-4 materials with plated-through-holes or clearance holes on the hard board backer area.

Flexible circuits were invented in the early 1950s, just a few short years after the development of the rigid printed wiring boards. They did not gain acceptance nearly as fast as the rigid board. One problem was that most of the materials available at that time were thermoplastic and could be used only at relatively low working temperatures.

Another concern was the need for flexible materials that would meet *both* the mechanical and electrical requirements and still yield producible circuits economically.

Flexible circuit substrates

The flexible circuit industry has come a long way since the early 1950s. Today, polyester, polyimide, fluorocarbon films, and aramid materials are used as substrates for flexible circuits. These dielectric materials are clad, with or without adhesive, with rolled-annealled copper or electro-deposited copper foil. Rolled-annealled copper is most widely used. Electro-deposited copper may be specified for applications where the finished circuit is not exposed to dynamic flexing.

Kapton and Pyralux polyimide substrates are the most popular, high performance flex circuit materials. They are often specified where all their superior properties are not required—much the same situation as FR-4 in rigid board laminates. DuPont supplies Kapton as base film to laminators of flexible materials and also supplies single- and double-sided copper-clad laminates under the trademark Pyralux. Adhesives used in laminations are proprietary formulations of acrylics, polyesters, epoxies, polyimides, and phenolics. In general, the acrylic adhesives offer the highest peel strength (8–15 lb) as compared to epoxy (5–10 lb) or polyester (3–6 lb).

Kapton polyimide film is specified where high performance is required. It can withstand repeated soldering at 500°F (260°C) and retains its excellent dimensional stability, chemical resistance, and electrical properties. It can also withstand multiple soldering and un-soldering operations that may be needed in field repair to replace faulty components. Polyester films such as Mylar have similar properties to Kapton, with the notable exception of heat resistance, which is 200°F max (93°C max). Polyester films are therefore not generally applicable for soldering applications. Soldering Mylar polyester film can be done (Polaroid did it in 1960s; see Fig. 12), but it requires more skilled, and thus more expensive, labor.

Fig. 12 Polaroid camera.

Aramid fibers such as Nomex can withstand temperatures up to 428°F (220°C) and offer chemical resistance and good electrical properties. They suffer from dimensional instability caused by variations in moisture content dependent on humidity.

Teflon fluorocarbon film is used in applications requiring chemical resistance and outstanding electrical properties, especially dielectric constant.

When selecting a dielectric substrate, consideration should be given to the thermal properties (both processing and operating), dimensional stability, the toughness of the film (both tear resistance and mechanical), electrical properties, flexibility, chemical resistance, flame retardance, adherability, and cost (see Table 3).

Table 3 Comparison of Flexible Materials

Dielectric material	Continuous service temp, °C	Tear resistance	Tensile strength, psi	Ultimate elongation, percent	Moisture absorption, percent	Dielectric constant, 10^3Hz	Dielectric strength, volts/mil	Dissipation factor, 10^3Hz
Polyimide	230	fair	25,000	70	2.9	3.5	7,300	0.0025
Polyester	150	fair	25,000	120	<0.8	3.2	7,500	0.005
Aramid paper	220	fair	10,500	8	<7.0	2.0	530	0.007
Fluoro-carbon	200	good	3,000	300	<0.01	2,1	2,100	<0.0001
Epoxy	100	good	3,500	20	0.8	5.0	500	0.010
ASTM test method used	—	D-1922	D-882	D-882	D-570	D-150	D-149	D-150

Source: Courtesy of *Insulation Circuits*, November, 1981

1.2 How to Interface with Flexible Circuit Manufacturers

The key to designing successful flexible circuit packaging is to start a dialogue with the supplier early in your design.

The best approach in solving a design problem is to establish realistic specifications and requirements for the proposed flexible circuit design.

First, become thoroughly familiar with the product requirements before starting the actual flexible circuit design. These should include electrical, functional, dimensional, assembly, and other requirements. Remember that the flexible circuit vendor cannot help you unless you can communicate your design criteria clearly. If your discussions might include some company trade secrets, it will be to your advantage to have the vendor sign a nondisclosure agreement. Remember, the vendor can only help if he fully understands the requirements of your design.

It is necessary for the design engineer to be familiar with and make reference to recommended flexible printed wiring design guidelines, such as the IPC Design Guide, and their impact on cost, reliability, manufacturability, etc. Don't overspecify your design. Keep your needs as simple as the product will allow.

It is important that the design engineer not delay the layout of the flexible printed circuit design. Design it at the same time as the basic product keeping flexible circuit guidelines in mind. The decisions made at this point have a direct impact on the price and reliability of the flexible circuit. This is the best time to establish contact with potential flexible circuit suppliers for design, specification, and process review. Most flexible printed circuit suppliers will provide this service and give comments and recommendations relating to functionality, producibility, and costs. The earlier this can be done in the design process, the better. Listen and pay attention to the suppliers, and benefit from their experience.

Now you are ready for the next important step—the prototype of your flexible circuit design. Prototype your flexible circuit design as early as possible. Definitely do your prototype early enough so that you can make adjustments and changes if necessary. You can learn a great deal from a few experimental prototype circuits assembled into your end product.

If any major problems are encountered at this stage, meet with your potential supplier's engineering personnel to discuss your design problems and to formulate a solution. This will also give you an opportunity to assess the vendor's capabilities in person.

After all major problems are identified and addressed at both ends (development and manufacturing engineering), give the selected vendor adequate time to prepare quotations, design and build tooling, generate production artwork, order special materials, process and submit tooling samples for testing and approval, and get into production mode before your high production volume requirements of parts are needed. Allowing enough time for these needed steps greatly decreases the likelihood of errors or oversights during the processing stage (see Fig. 13).

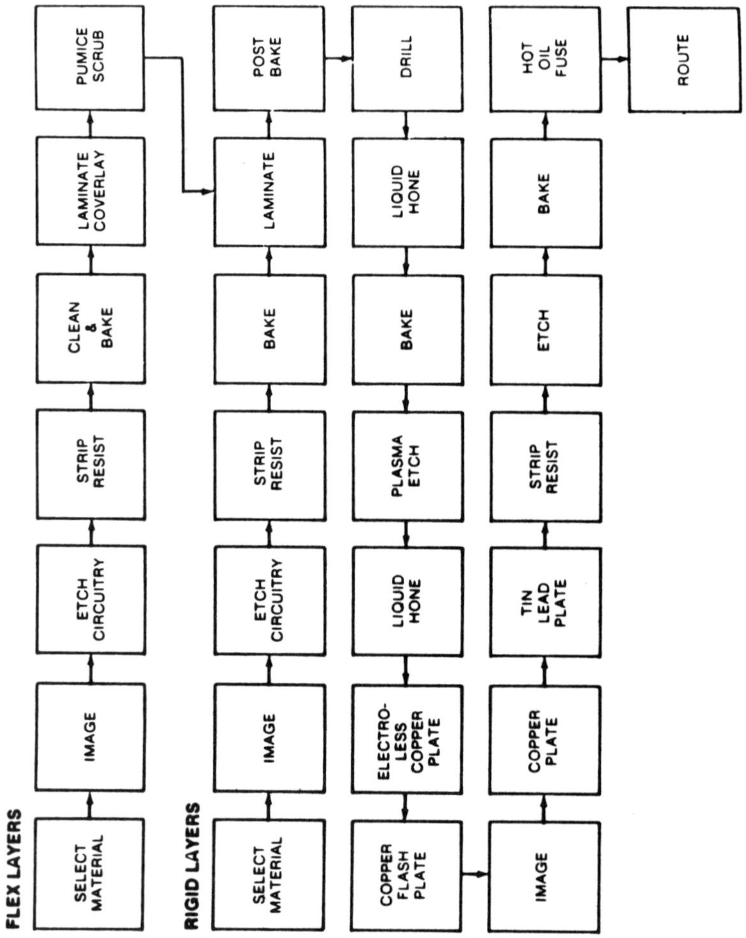

Fig. 13 Rigid-flex multilayer printed circuit board flow cycle.

1.3 How to Price Flexible Circuit Design

Although flexible printed circuits typically are used to optimize space and to minimize assembly restrictions, they are often specified where cost must be kept to a minimum.

Flexible printed circuits may cost more to manufacture per square inch than conventional printed wiring boards. So, desired cost reductions are realized mostly through assembly costs. These cost savings can be substantial.

Wave soldering, automatic SMD placement, and other cost-saving assembly methods can all contribute to assembly cost reductions. However, these potential savings can disappear if dimensional tolerances are too tight. Material costs can vary widely. It is easy to understand that design engineers must be knowledgeable about the design criteria, materials, tooling, SMD assembly, and processes used in flexible circuit manufacturing.

Flexible circuit design criteria

The following do's and don't's should be kept in mind when designing and laying out a flexible printed circuit.

Terminal areas of pads

Ratios less than 2 to 1 in size versus hole diameter will present manufacturing yield problems as the numbers of base layers increase (see Fig. 14).

Conductor sizes

The conductor size will be dictated by using the specified current requirements drawn from the monograph (Fig. 15). However, the circuit layout must also be considered. Spac-

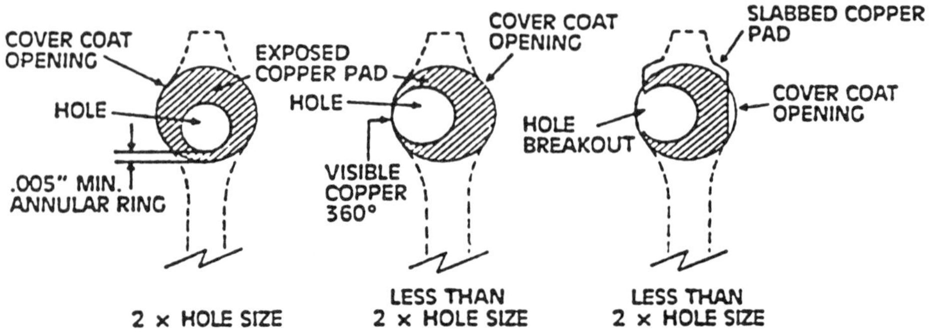

Fig. 14 Size versus hole diameter.

ing is also a function of the specification, and you should keep to the maximum where possible. Another important point is that conductor runs should be evenly distributed to avoid unnecessary open areas or congested layouts.

Hole-to-pad and hole-to-pin diameter ratios

Hole-to-pad ratios should not be less than 2 to 1 to assure consistent and producible annular ring requirements. The hole-to-pin diameter ratio should allow for a 5–7 mils clearance between the maximum pin diameter and the minimum finished hole size (see Fig. 16).

Edge-to-conductor dimensioning

The dimension from the edge of a circuit to a conductor trace or pad should be a minimum of 30 mils and allow for ± 15 mils per edge for outline tolerances, material misregistrations, and additional process tolerance build-ups (see Fig. 17).

(USE IN DETERMINING CURRENT CAPACITY AND SIZES OF ETCHED COPPER CONDUCTORS FOR VARIOUS TEMPERATURE RISES ABOVE AMBIENT)

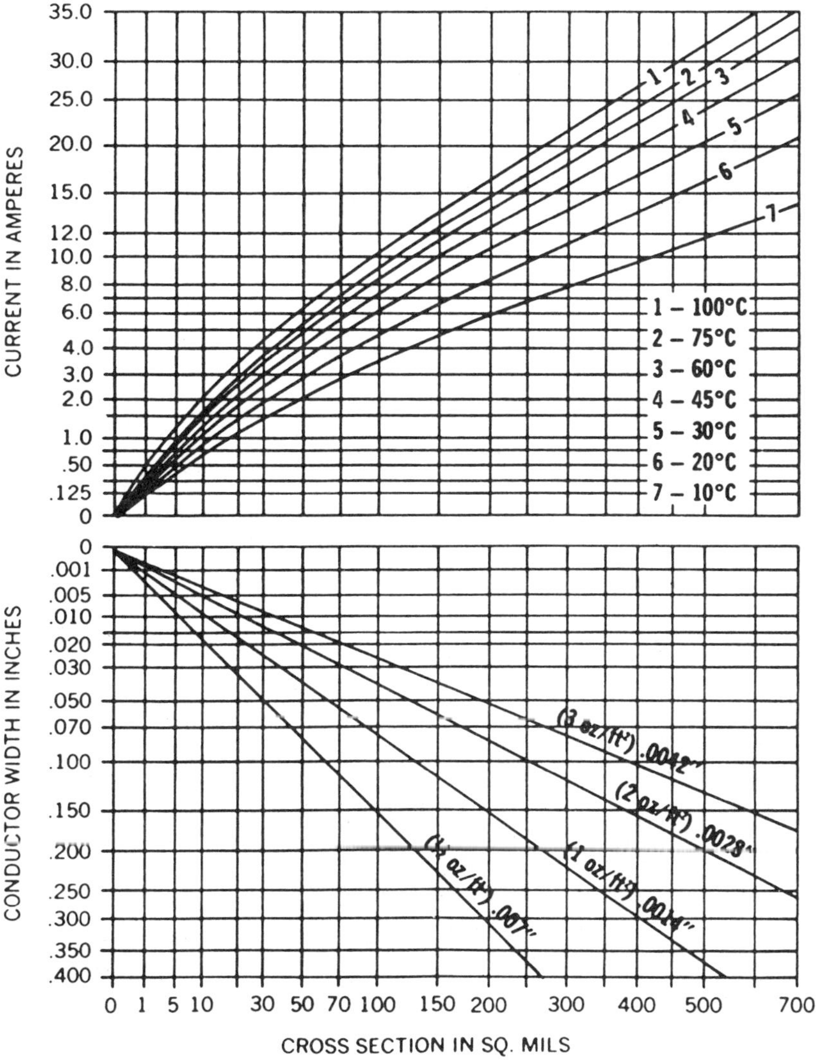

Fig. 15 Graphs for use in determining current capacity and sizes of etched copper conductors for various temperature rises above ambient.

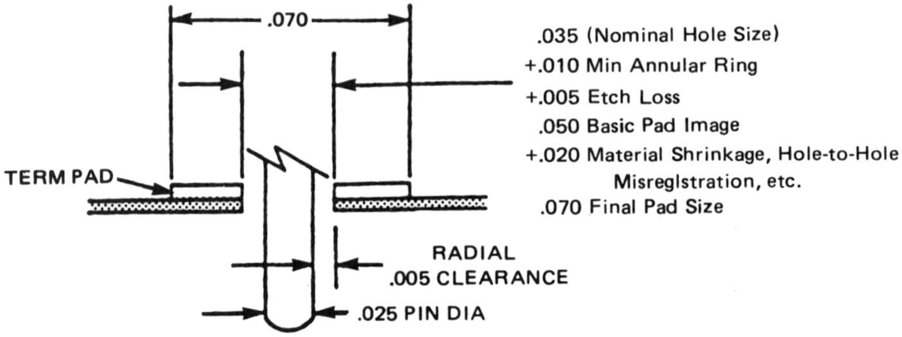

Fig. 16 Hole-to-pin diameter ratio.

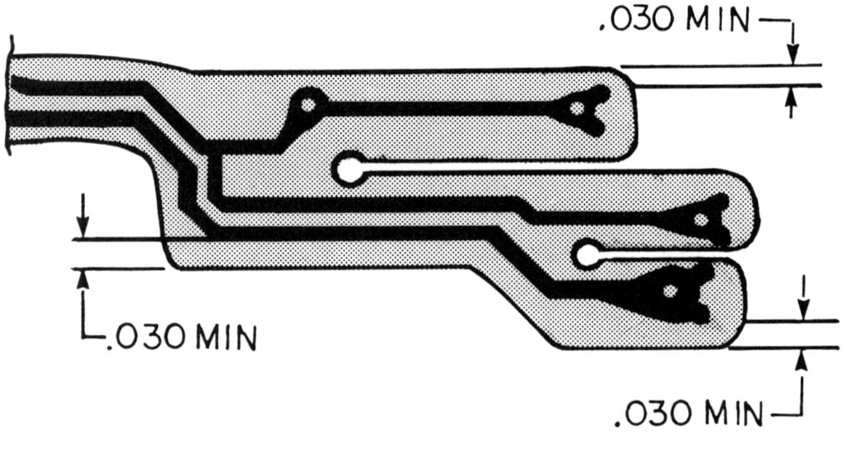

Fig. 17 Edge-to-conductor dimensioning.

Bending and folding

Designs should avoid I beam effect with conductors or crossovers stacked instead of staggered in the circuit bend areas. A single-sided rolled-annealled copper circuit with a radius equal to the material thickness can be folded on itself three cycles before failure. A double-sided circuit of similar construction should have a radius of ten to twelve times the material thickness (see Figs. 18 and 19). It should be noted that flexible circuits of Kapton polyimide film have been designed and tested to 500,000,000 cycles without loss in circuit performance. Flexibility can be designed into circuits when requirements are known.

Tear resistance

Although most flexible circuit substrates are inherently tough, it is standard flexible circuit design practice to allow for the largest possible radius at all internal corners. It is also good practice to include glass-cloth reinforcement in the laminate design at internal corners for added reinforcement (see Fig. 20).

Cover coat openings

If pads cannot be laid out with tabs and fillets, then the cover coat should lap onto the pad area for a minimum of 5 mils for a circumference of 270° (see Fig. 21).

Dimensions and tolerances

Because of the inconsistencies in hole diameters, conductor spacing and size, hole-to-pad and hole-to-pin rations, edge outline tolerances, and registration, the dimensioning and tolerances applied to the design, layout, and manufacture of flexible circuits must be

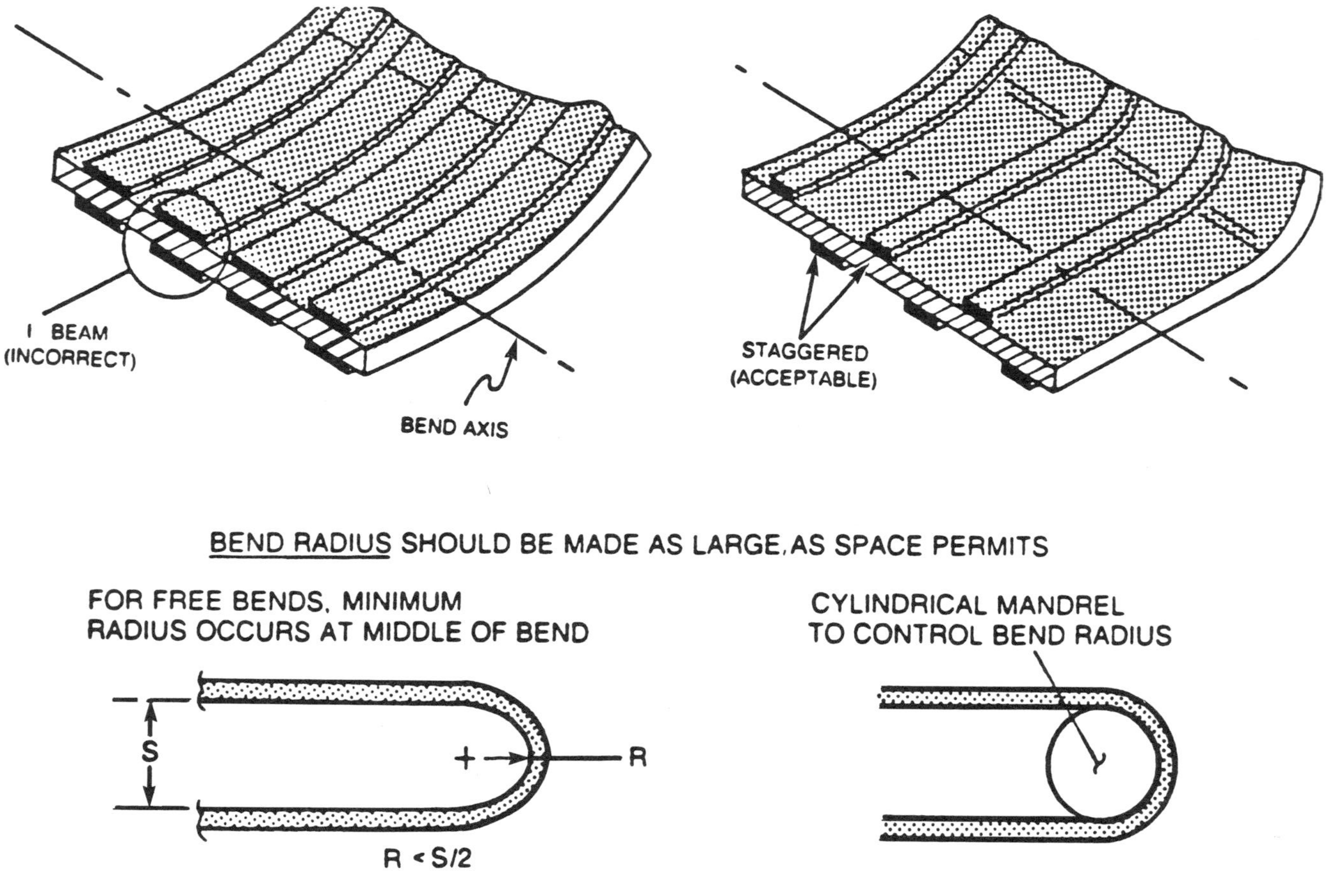

Fig. 18 Illustrations of bend radius.

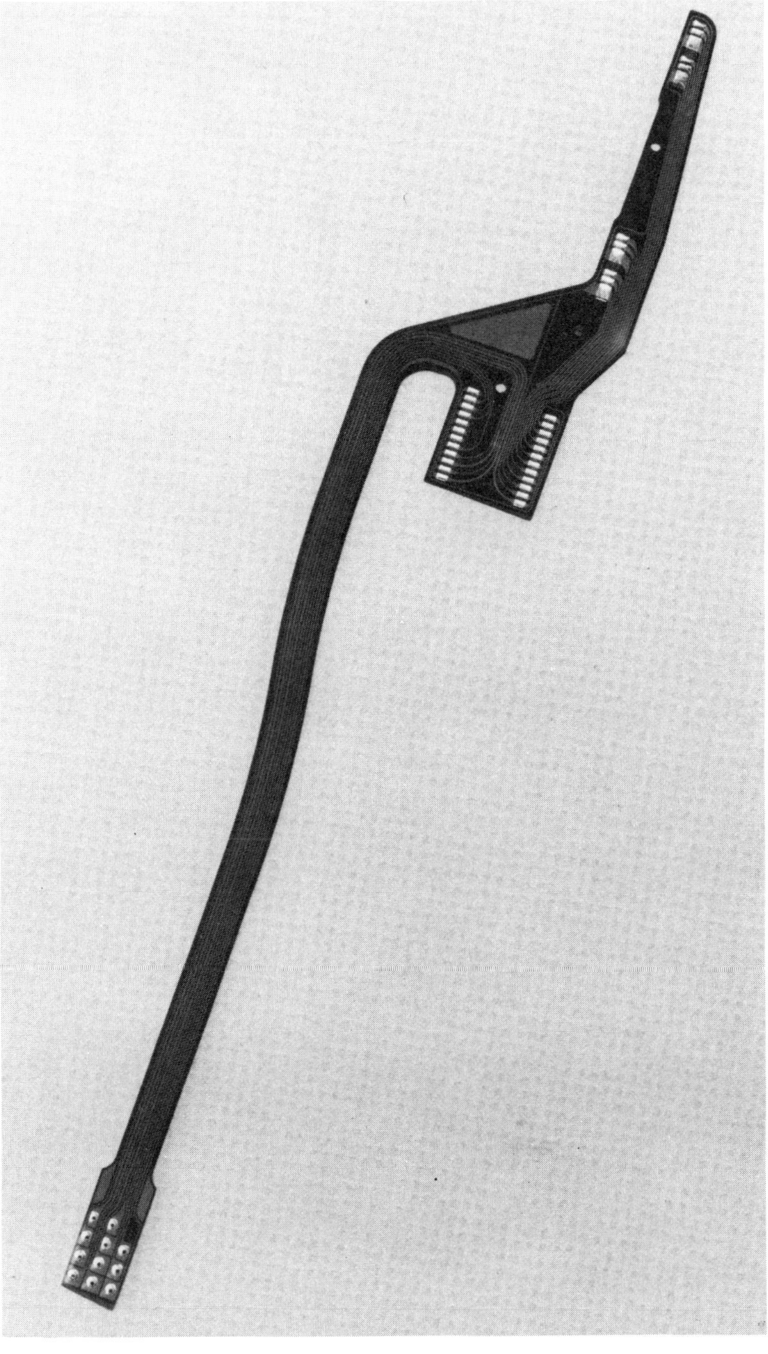

Fig. 19 IBM computer disk drive circuit.

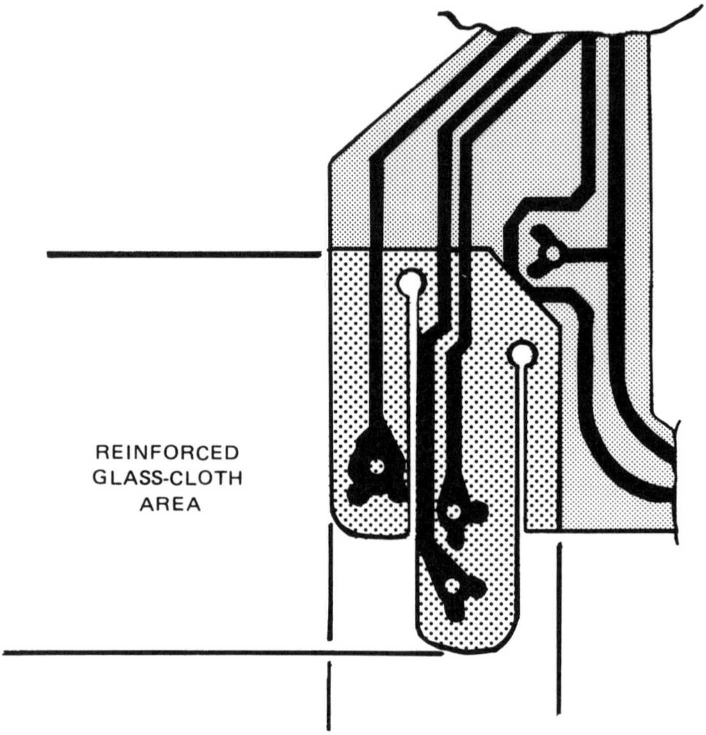

Fig. 20 Glass-cloth reinforcement in the laminate.

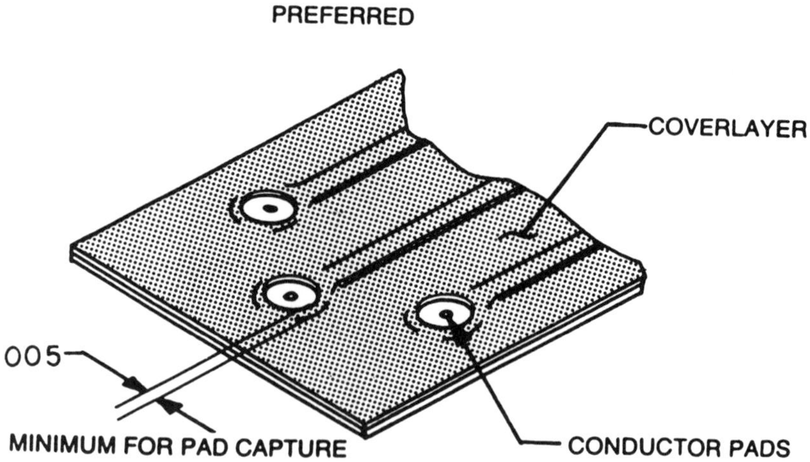

Fig. 21 Preferred cover coat openings.

practical so that present-day manufacturing techniques can be used.

Now that the basic engineering and material selection has been made, the next decision for the design engineer is to determine if a short production run will be needed, or if a full-scale production run is called for to minimize the unit cost.

2
PACKAGING CONCEPT: A CASE STUDY

2.1 The Advantages of Volume Design over Area Design

Portable, two-way communication equipment provides a useful application for printed wiring in either flexible or rigid substrate materials.

A basic radio printed wiring board can be designed to retain and interconnect the various required electronic components for the transmitter as well as the receiver elements, all within minimum space requirements for optimal efficiency.

High circuit density has been achieved in the past through the use of closely placed discrete components or subcomponents on a printed wiring board constructed with the use of multilayer techniques. These printed wiring boards can then be stacked in a common housing frame.

A disadvantage of the present printed wiring board approach is that the chassis board, once designed and laid out, is fixed operationally and not readily adaptable to handle additional or alternative operational features. A different printed wiring board for each combination of operational capabilities will increase cost unnecessarily, is inefficient and inflexible, and may impair reliability of the product.

There are many instances where available space is at a premium due to miniaturization, portability, and the relative density of the electronic components. Utilization of available space becomes especially critical when the design uses both components with leads and leadless devices on the same printed wiring board. This printed wiring board will be finished into a module to be packaged in an item of associated electronic equipment. The need to save weight and space, reduce costs, and increase efficiency, flexibility, and reliability led to the development of the Flex-O-Pak concept, which is a packaging technique based on volume design.

Prior to the development of the Flex-O-Pak technique, the state-of-the-art technology used by communication development engineering was mostly based on small hybrid modules and printed wiring boards. Two types of printed wiring boards were used.

The printed circuit board used to interface with the main radio board is called the "*mother board.*" This board was approximately .890 in. in length, 2.50 in. in width, and .062 in. thick. The material used was FR-4, with two ounces of copper clad on both sides (see Fig. 1).

The mother board was physically attached to the main radio board using an interconnecting flex circuit. The flex circuit function was to pick up the selected signals as well as the power and ground connections. This data was then interfaced with the preassigned socket pins in the mother board (see Fig. 2).

The printed wiring board with the optional electronic circuitry is called the *daughter board*. This board was made of .031 in. thick FR-4 with two ounces copper clad both sides. The duaghter board measured approximately .95 in. in height and about 2.30 in. in width.

Due to the available surface area on the mother board, and the size of the electronic components, only three daughter boards could be used in this design. The remaining critical circuits were designed into two hybrid mod-

Fig. 1 The mother board. Courtesy of Motorola, Inc. Communication Division. Copyright © Motorola, Inc.

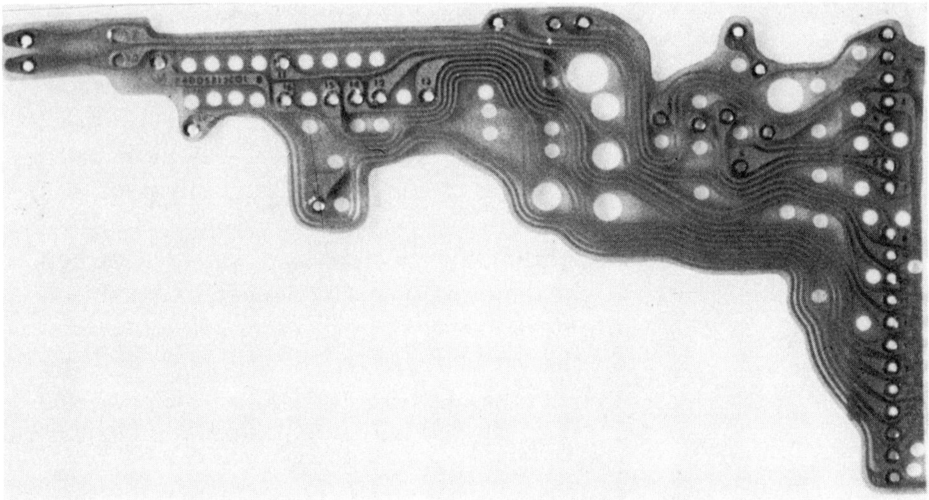

Fig. 2 Illustration of socket pins in mother board. Courtesy of Motorola, Inc. Communication Division. Copyright © Motorola, Inc.

ules. The density of the daughter boards is readily apparent in Fig. 3.

Notice that this package scheme uses the smallest available leaded resistor. The other semiconductors were specified in the smallest packages available from commercial vendors. The transistor package is a mini-version of the TO-92 package. The ICs are a mixture of custom design ICs and over-the-counter devices.

However, the IC mounting technique is different from the soldering process intended for the flatpak device. The IC flatpak leads were fixtured to bend downward like in a dual in-line package. The ICs were then inserted into plated-through-holes in the daugh-

Fig. 3 The daughter board. Courtesy of Motorola, Inc. Communication Division. Copyright © Motorola, Inc.

ter board similar to conventional dual in-line devices. The printed circuit daughter board assembly was then soldered on a conventional wave solder machine. The right angle connector shown provided the interface between the daughter boards and the preassigned socket pin assembly of the mother board.

As previously mentioned, it was necessary to design two hybrid modules in addition to the three daughter boards. These two hybrid modules were approximately .15 in. thick, .90 in. in height and .89 in. in width.

The biggest drawback of the daughter board and hybrid module package was individual testing of the unassembled parts prior to insertion into the main radio board sockets. Each daughter board had to be tested separately, as well as the two hybrid modules.

After the individual test, the complete assembled system had to be tested again.

The daughter board and hybrid module package did make it possible to add optional circuitry to the basic radio, but the material costs, labor costs, and redundant testing made the manufacturing cost very high and less desirable for a high-volume production run.

The basic radio design also evolved from an eight channel radio to a fully synthesized sixty-four channel radio.

Although the radio frame remained the same size, the optional circuitry space was reduced by 50%. Where we had about $.18^3$ before, the new radio design allowed only 0.9^3 for optional circuitry.

A printed circuit board area study was conducted and the results were catastrophic. The new mother baord area was reduced to .40 in. in length by 2.50 in. in width, with the board thickness remaining .062 in.

The remaining space was equivalent to two of the spaces occupied by the daughter boards with the individual daughter board size measuring .95 in. in height, 2.40 in. in width, and .031 in. in thickness.

The .100 in. pin centers right angle connector limited the I/O in the two daughter boards to about forty pins. Thirty-four of those available interconnecting pins were needed to interface with the main radio board signals, leaving only six pins remaining to interface from one daughter board to the other.

Due to the new circuit density, the limited number of I/O pins, and the augmented material parts count, it was evident that the two daughter boards approach could not be implemented.

Clearly a new space saving approach was needed.

The possibility of designing two larger hybrid modules to house the optional design was carefully reviewed. However, the bad yield of the previous hybrid modules, and the crystal package size, made this approach much less desirable. The hybrid approach also did not solve the interconnecting problem, or the inability to interface with each other outside of the limited number of sockets available in the mother board.

At this time, a packaging revolution was gathering strength in the United States. A few years earlier in Japan, electronics manufacturers first began to attach miniaturized components such as chip resistors and capacitors directly onto the printed wiring board surfaces, an automated technique that evolved from thick-film hybrids.

To meet this need, electronic component manufacturers had created whole new families of tiny active, passive, and semiconductor devices to fill the new demand for SMDs.

Because of these new developments and past experience with flex circuit designs, management began to contemplate the idea of using flex circuits with surface-mounted components to design the new option package. If the circuit layout could follow the schematic I/O, the interfacing of the electronic functions would require a minimum number of interconnects.

A preliminary bill of materials and a written proposal plan were presented to management. Since this approach was considered a new concept, it required a comparison between the two previously proposed packages, plus the new flexible substrate package.

The design required a pluggable module, with interconnecting pins at the edge of the flexible circuit. There were currently only two methods to affix pins onto a flex circuit design.

Sculptured interconnects

Sculptured interconnect flex circuits are produced in panels of 10 mil thick copper. This base copper is reduced in thickness to 4 mils for conductors, and the connecting pin areas are left at the full 10 mils thickness for strength and rigidity (see Fig. 4). The limitations of this method are:

1. Only single-sided flex circuits can be made this way.
2. The maximum number of insertion cycles is about ten per given set of pins.

Although this technique was not practical for this application, it is one of the best and simplest ways to interface between a flex circuit and rigid board. Advanced Circuit Technology in Nashua, New Hampshire, holds the patent for this technique, and Interconics Company, Nashua, New Hampshire, is licensed to do similar type work.

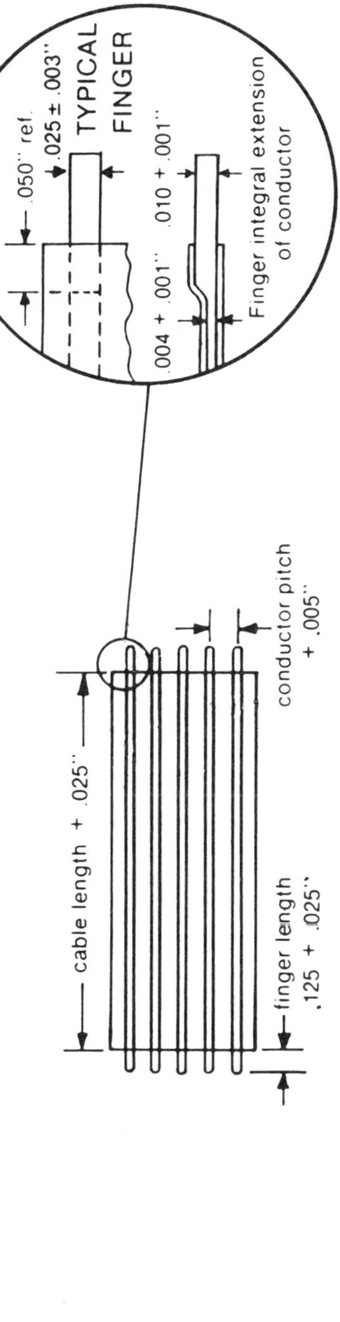

Fig. 4 General specification for a sculptured interconnect. Courtesy of ACT, Nashua, New Hampshire.

Multicrimps terminations

The second available method was the multicrimp termination. Multiple crimp contacts rely to some degree on accepted and proven technology, especially as far as materials and plating are concerned. The contact's barrel shares the common characteristics of four separate tines which pierce the insulation and trap the conductor to establish a residual force system with multiple redundancy (see Fig. 5).

Although several design changes in these crimp contacts have been made, reliability, the height requirement, and the single-sided design restriction, made it incompatible with the design criteria. AMP Inc., Harrisburg, Pennsylvania, has been developing tests to establish the overall validity of the crimp-based connection system.

The solution to the design problem was

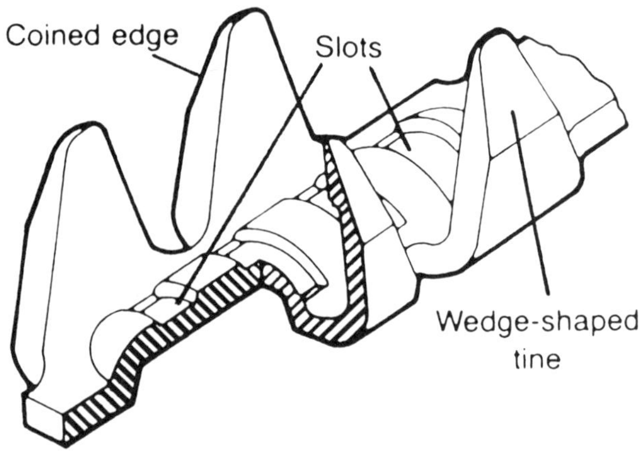

Fig. 5 Four-tined multiple crimp contact. Courtesy of Electronic Packaging and Production, March, 1984.

not readily available in the industry. However, we developed a process that will allow the affixing of interconnecting pins directly onto a flex circuit pad.

This task was achieved with the help of a high temperature solder process, which will be discussed in Chap. 4.

An important design consideration in this packaging technique is the separation of the leaded components from the SMDs. This is necessary because we want to be able to trim the leads as close to the flexible substrate surface as possible. This could not be done if leaded parts and surface mount devices are found on the same plane, at opposite sides from each other. If both leaded components and SMDs are found on the same plane, the closest you can trim the leads is the height of the highest leadless device plus a minimum clearance of 15 mils above the top of the SMD.

Keeping in mind the available volume area, the best method of supporting the leaded components on the flexible substrate was selected. All SMDs were placed on the two extreme ends of the flexible substrate. This included most of the resistors, capacitors, transistors, and diodes.

The center area of the flexible substrate was laminated to a 15 mils thick non-copper clad FR-4 epoxy glass rigid board panel. This area housed all the modified IC flat-paks and the unique leaded components such as the crystal, electrolytic capacitors, zener diodes, and some special resistors. This rigid board panel provided support and prevented mechanical stress on the leaded components.

The rigid board panel area extending beyond the flexible substrate was used for handling the assemblies during the affixing of the SMDs by the automatic placement machine.

Apertures were provided on the rigid board panel for the component leads to pass through to the plated-through-holes of the flexible substrate. The IC apertures on the rigid board panel were slotted to facilitate insertion of the close distance IC leads onto the flexible substrate holes.

In order to optimize the available rigid board panel area, two flexible substrates were used per panel.

Since the solder side of the flexible substrate was mostly populated with SMDs (about 90 parts), this side was first in the assembly process. The leadless devices were attached and bonded to the flexible substrate surface.

The next step was the manual insertion of all the leaded components. Since a low profile was needed, the parts were glued in place to prevent the components from rising as the flexible substrate assembly went over the dual wave solder machine.

Figures 6–8 show the flexible substrate as supplied by the vendor, showing both assembly processes.

The circuit I/O were evenly divided between the two flexible substrate edges. The total number of I/Os required was limited to thirty-eight connections. Four of those connections were ground returns, allowing the mixing of digital and analog logics in the same flex circuit layout without the internal looping of the two different circuit ground returns.

The ground pin connections were lat-

Fig. 6 Flex-O-Pak as supplied by vendor. Courtesy of Motorola, Inc. Communication Division. Copyright © Motorola, Inc.

er soldered onto the corresponding housing ground tabs, providing the proper way to ground the two different circuits, *and* creating EMI shielding to protect the flexible substrate module from interference due to the radio RF signals (see Fig. 9).

Fig. 7 Flex-O-Pak leadless assembly. Courtesy of Motorola, Inc. Communication Division. Copyright © Motorola, Inc.

Cost savings in testing

The old optional package design required that all three daughter boards be individually tested as well as the two hybrid modules: a total of *five* individual tests. One of the advantages of the flexible substrate package approach is that the entire self-contained module is tested once after the flexible substrate assembly is cleaned and the component leads properly trimmed.

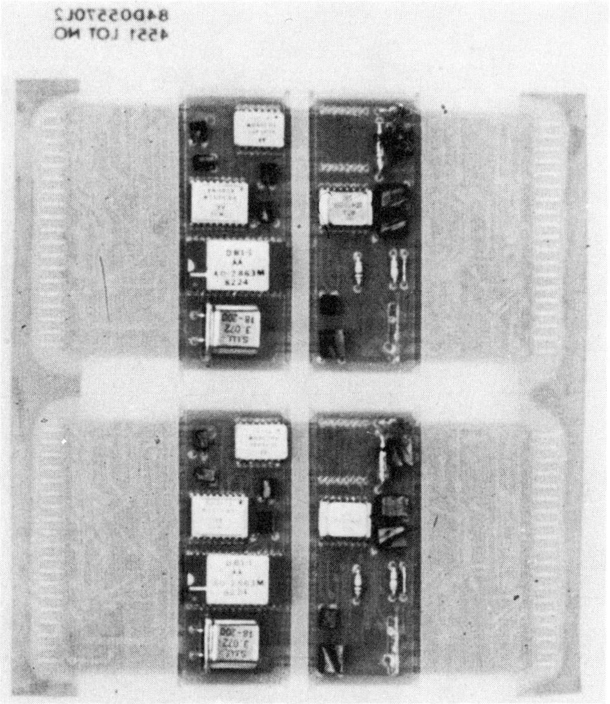

Fig. 8 Flex-O-Pak leaded assembly. Courtesy of Motorola, Inc. Communication Division. Copyright © Motorola, Inc.

The Flex-O-Pak test fixture runs a software control program which allows the assembly line operator to determine if any of the indicated faults are due to component failure or a bad solder joint. Next, the operator folds the tested flexible substrate assembly in an accordion-like manner and pushes the interconnect pins through the aperture in the plastic header. Insertion of the flexible

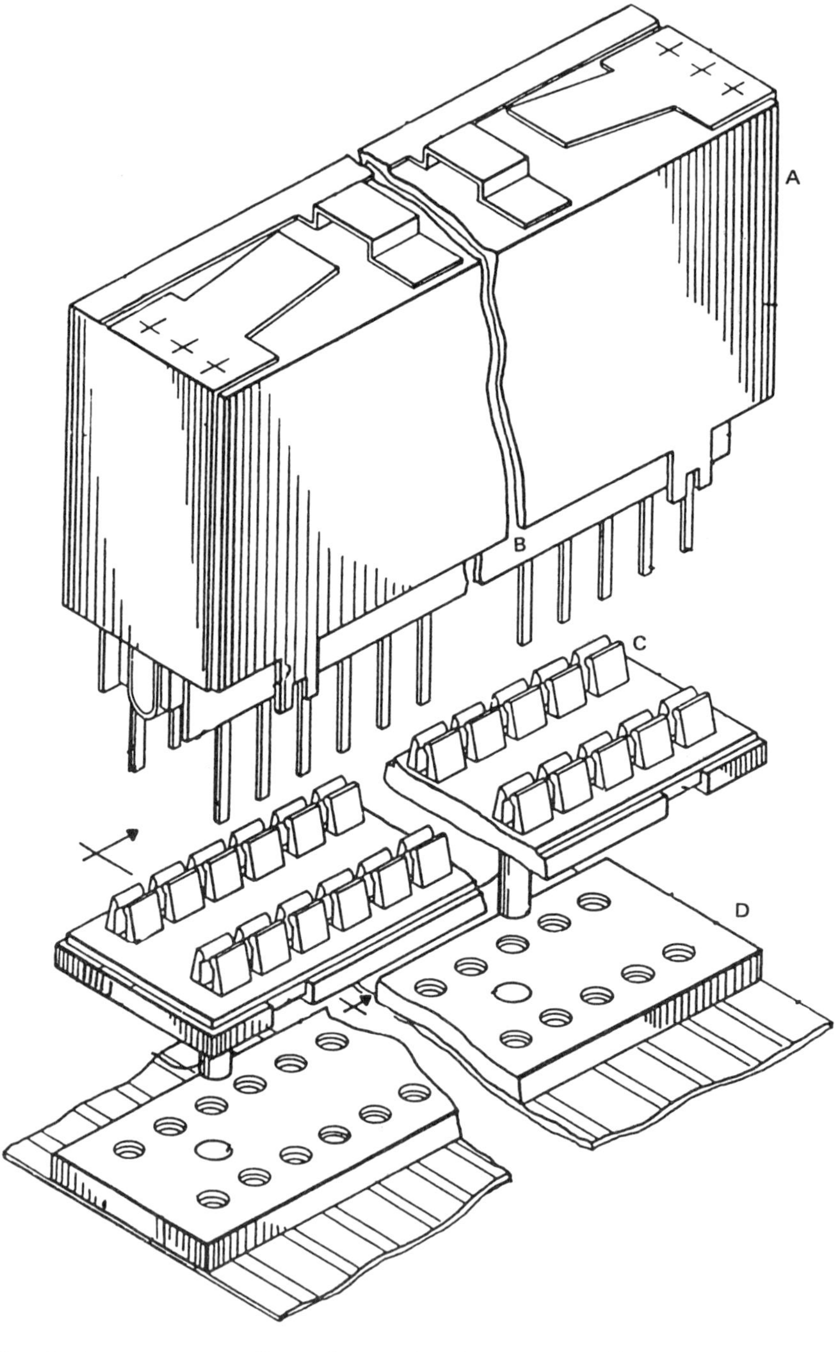

Fig. 9 Proper grounding of two circuits. A, metal housing; B, Flex-O-Pak; C, plastic header; D, interconnecting flexible circuit.

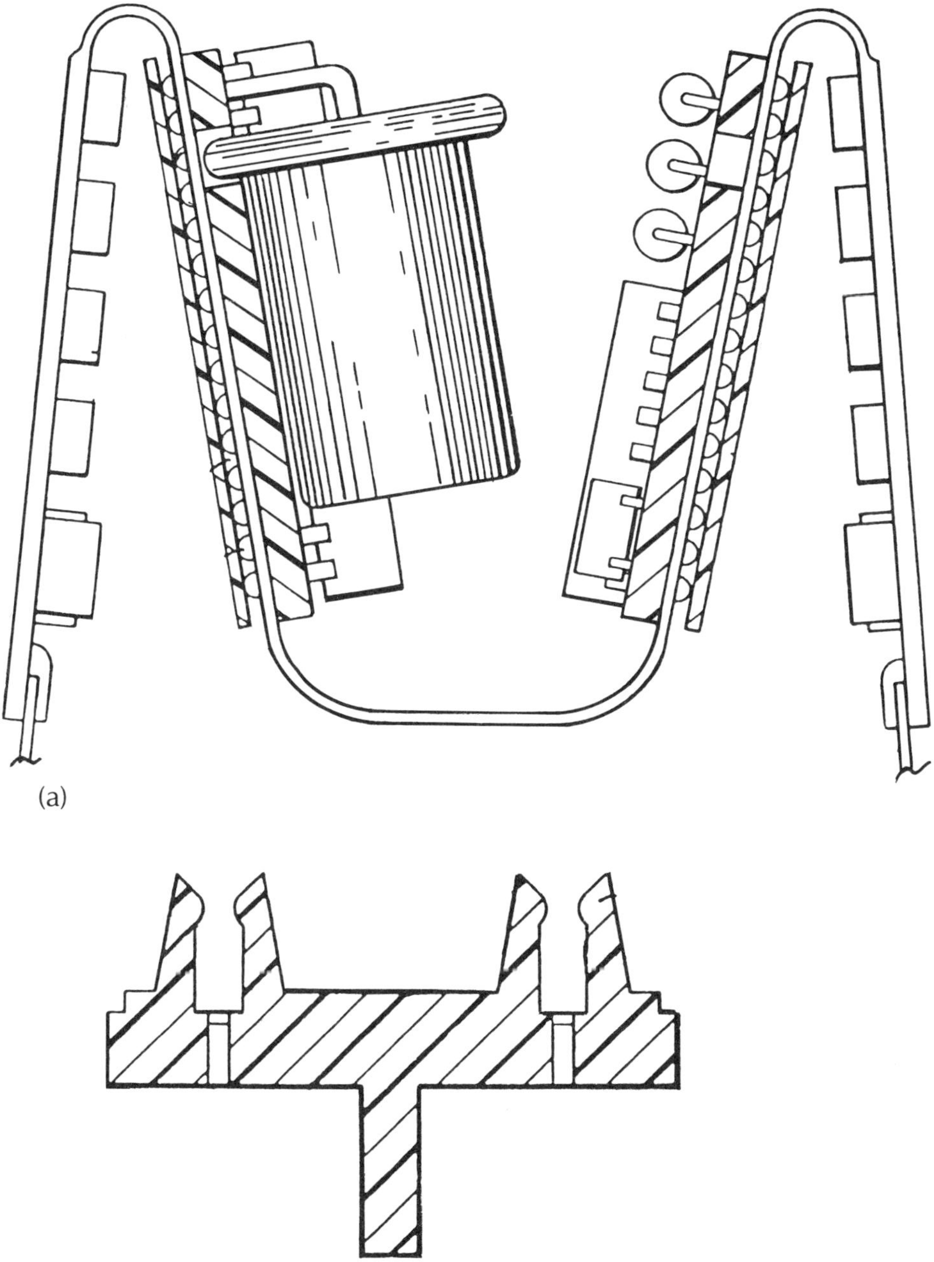

Fig. 10 (a) Flex-O-Pak folded in an accordian-like manner. (b) Plastic header cross-section.

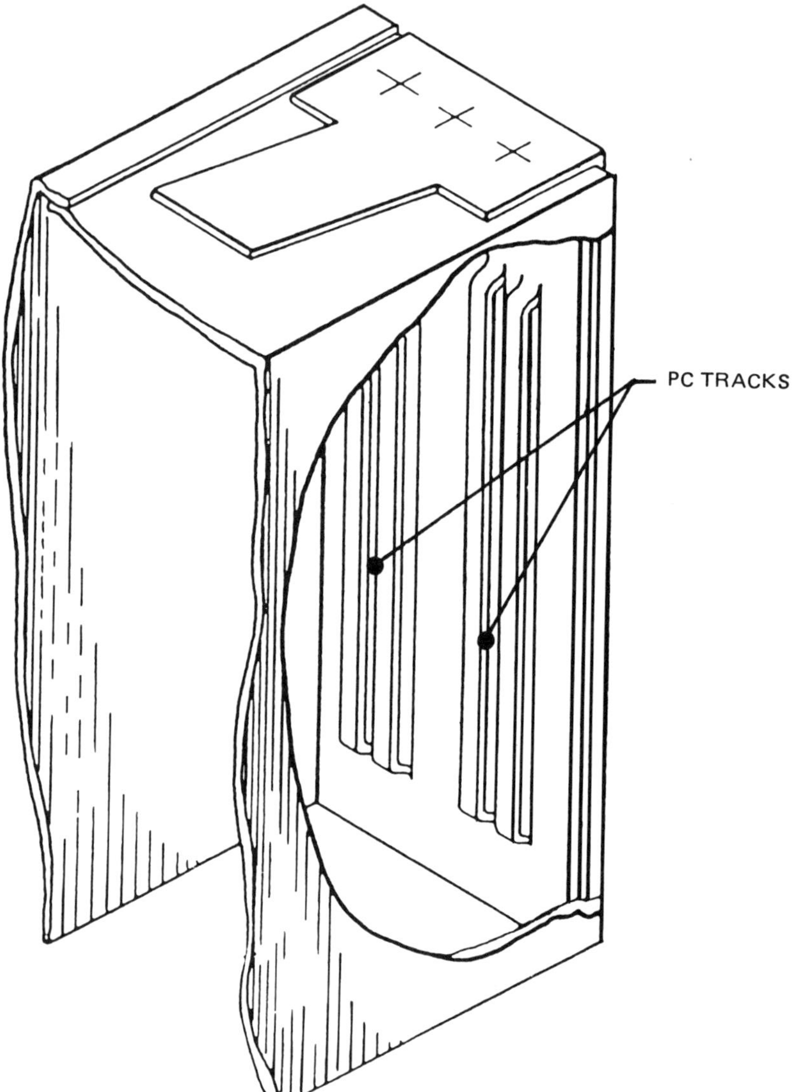

Fig. 11 Housing cross-section.

substrate assembly into the housing inner tracks in the next to the last step (see Figs. 10 and 11).

One or more electrical tests are performed on the flexible substrate to check for possible shorts. If none are found, the housing ground tabs are bent onto the ground pins and soldered.

The Flex-O-Pak design technique has also made repair and assembly less costly. Repair of the flex circuit assembly is now possible at a fraction of the cost of replacing a hybrid module. The flexible pyralux substrate can be repaired using a controlled-tip soldering iron. The plastic header and metal housing are expendable due to the low cost.

Chap. 3 discusses important considerations for design of a successful flexible substrate.

3
SMDs ON FLEXIBLE CIRCUITS

3.1 Design Considerations for Leadless Devices

The rapidly increasing use of SMDs on flexible substrates has made it necessary that some guidelines be established by the industry in this relatively unfamiliar area. Our experiences in the design of a fairly complex double-sided flex circuit layout employing SMDs, have pointed out some design rules which are critical when using SMDs on a flex circuit layout.

Unlike the affixing of SMDs on rigid printed wiring boards, the flexible substrate offers several challenges to the design engineer of a flexible circuit layout.

The cover coat material

The first, and most important, challenge deals with the flexible substrate cover coat material. The flex circuit cover coat material performs two functions:

1. It serves as a solder mask similar to its use on a rigid printed wiring board.
2. It adds depth to the flex circuit design.

The thickness of the flex circuit cover coat material plays an important role in the flexible substrate design. The average cover coat material thickness, including an adhesive layer, measures between 3 and 5 mils. This cover coat spacer keeps the SMD terminal barrier 3-5 mils away from the flexible substrate copper pad surface. Because of this design difference peculiar to this flex circuit construction process, the flex circuit vendor should deposit as much tin/lead plating as possible on the exposed copper pads (see Fig. 1).

The second design consideration is the texture of the flexible substrate cover coat material. Since the selected flex circuit substrate is Kapton polyimide film, it is recommended that the cover coat material surface be pumiced to remove the smooth shining surface. It should then be thoroughly cleaned prior to the solder reflow of all the exposed copper pads. This creates a better bonding surface between the adhesive dot used to attach the SMD and the flexible substrate surface. Also, it will minimize the adhesive spreading into other adjacent adhesive dots.

The third and last design consideration deals with the flexible substrate's bendability. It is recommended that the cover coat material extend the entire length of the flexible

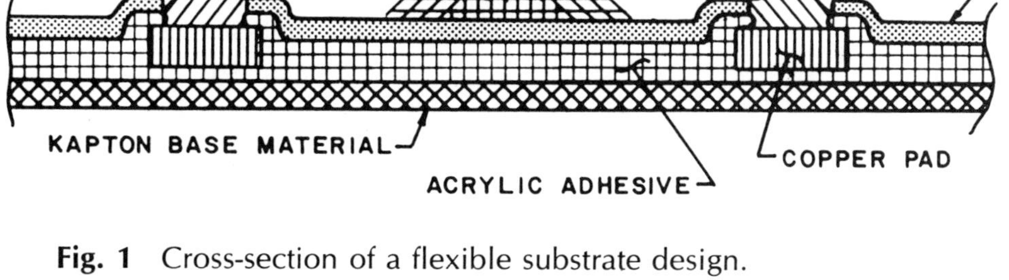

Fig. 1 Cross-section of a flexible substrate design.

substrate on the solder side. On the far (component) side of the flexible substrate, the cover coat material should extend only over the printed copper traces, leaving all bend areas free of cover coat material where there are no traces (see Fig. 2).

The cover coat opening

The opening in the cover coat material determines how much solderable pad area is available. An aperture measuring the same width as the leadless device will be a good rule to follow. In reality, this method provides an opening approximately 10 mils less than the width of the part because the adhesive used to laminate the cover coat to the base material will flow, creating a ring within the cover coat material opening (see Fig. 3).

To determine the space needed between the SMD terminals, it is recommended that the exposed area of the cover coat material measure 240° at both ends of the leadless component (Fig. 4). This assembly technique will offer the maximum latitude for affixing the SMDs with a dual-wave solder machine.

Notice that the leadless component pad area is square and the cover coat material aperture is round. There are two reasons for this which complement each other.

The first is the need to secure all the copper pads to the base material. This is usually done by adding extra tabs to the round pad so that the cover coat aperture limits the copper pad opening to a small area, preventing the copper pad from lifting off the flexible substrate surface. The second is the cost advantage of being able to drill regular holes

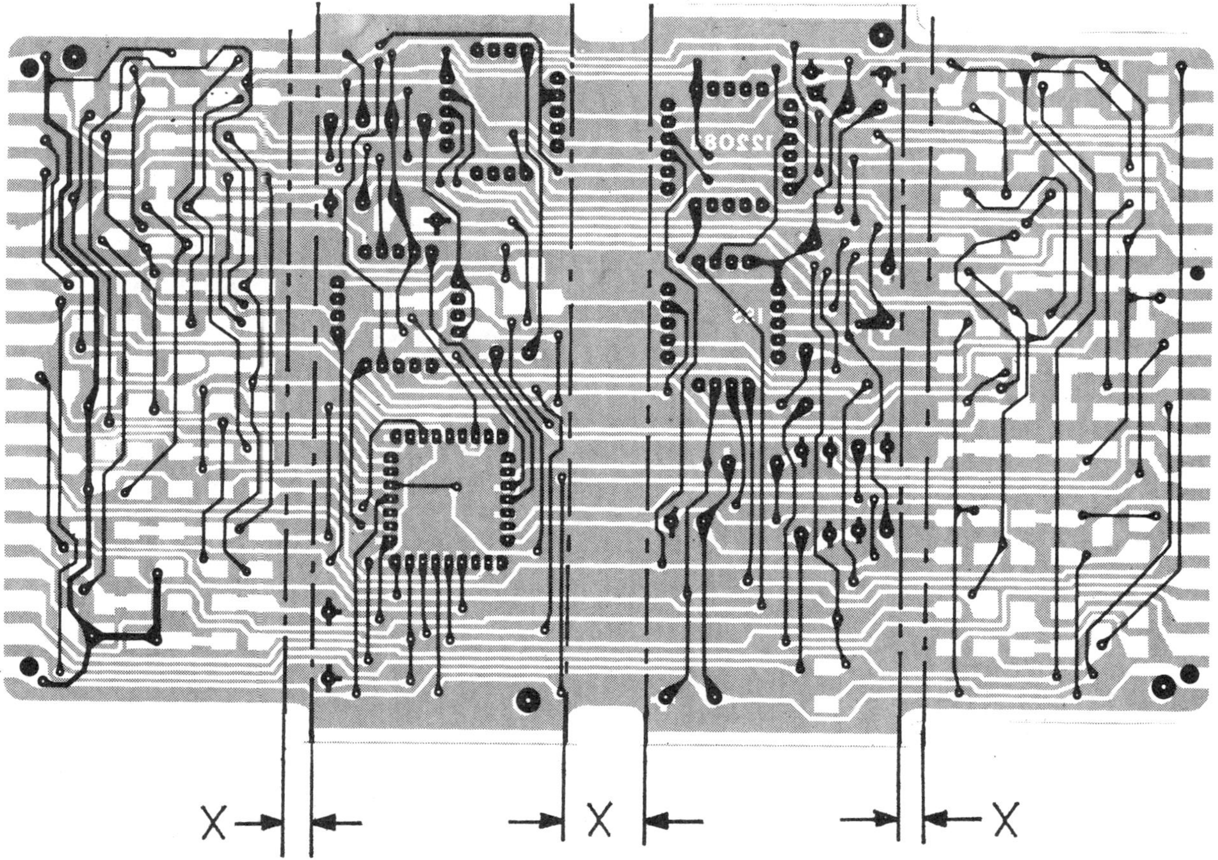

Fig. 2 Cover coat material applied only over the printed copper traces. X-Denotes no cover coat area. Copyright © Motorola, Inc.

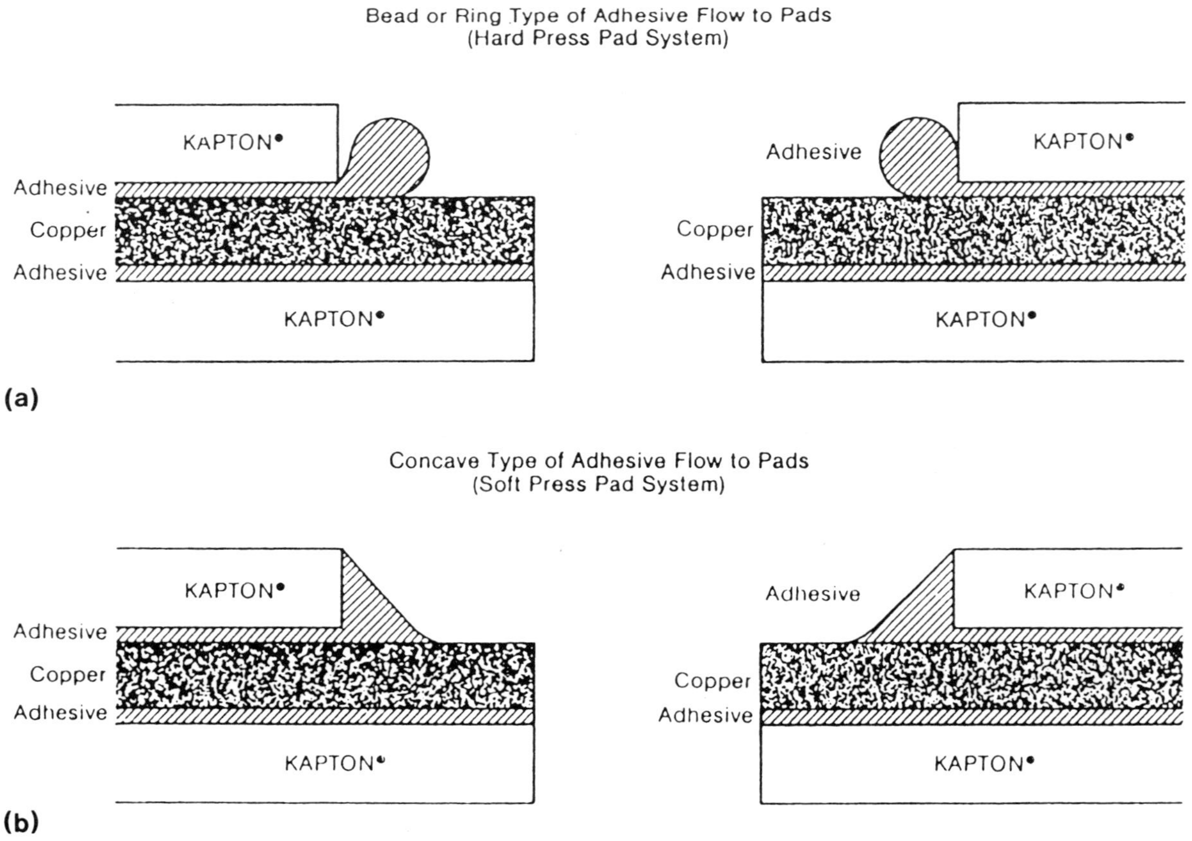

Fig. 3 Adhesive flow to the pads. (a) Bead or ring type of adhesive flow to pads (hard press pad system); (b) Concave type of adhesive flow to pads (soft press pad system).

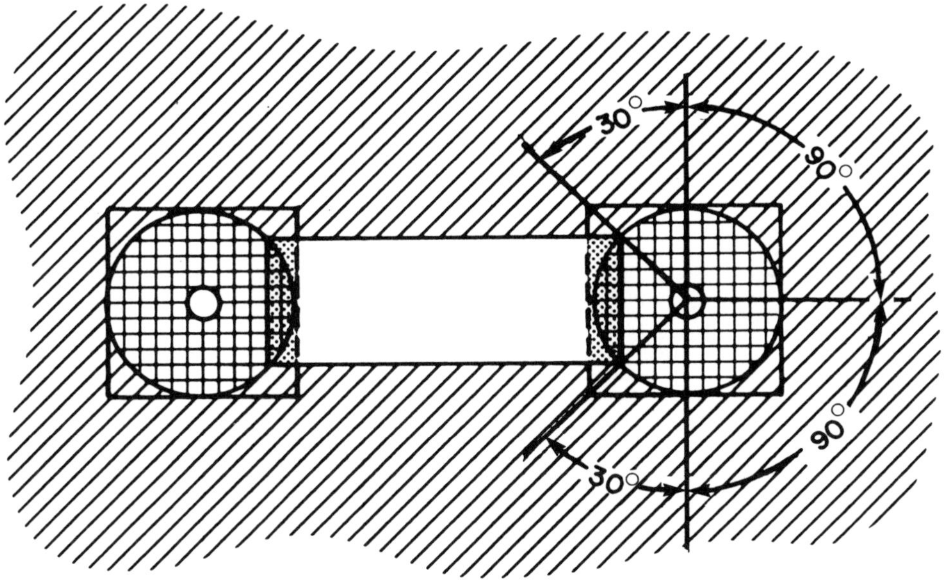

Fig. 4 Measurement of cover coat material shaving 240° of exposed area.

compared with the high cost of using punching tools.

The SOT-23 pad layout is not too different from the discrete components pad pattern. Figure 5 shows the ROHM CORP.'s SOT-23 pad layout. Although the terminal lead measures approximately 15 mils by 15 mils, it is recommended that a 50 mils square pad with a 40 mils diameter aperture be used. This space pattern will provide the necessary assembly latitude for automatic component placement.

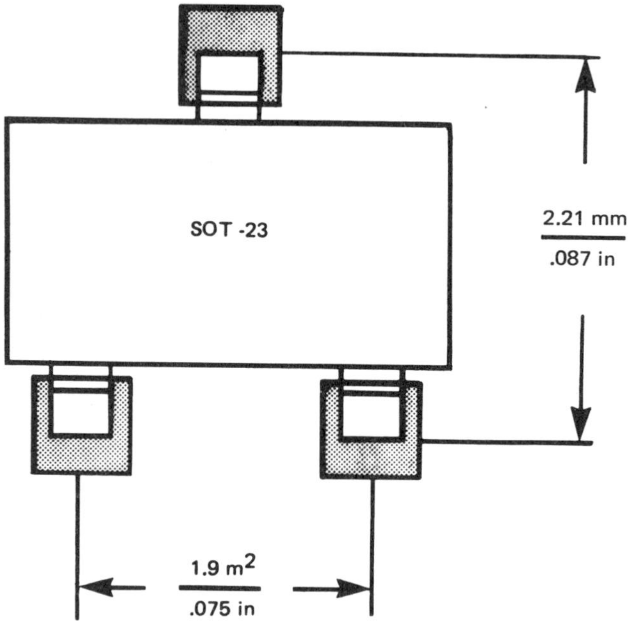

Fig. 5 The SOT-23 pad layout.

The substrate

The use of a flexible base material requires that provision be made for internal tooling holes in the design layout (Fig. 6). These internal tooling apertures are important in affixing of the flexible substrate to a rigid board panel (if needed), and provide the datum point for programming the automatic parts placement machine.

Artwork masters

Although the flex circuit layout and the art work master for the "Flex-O-Pak" were generated manually, there are some CAD/CAM systems capable of providing computer gen-

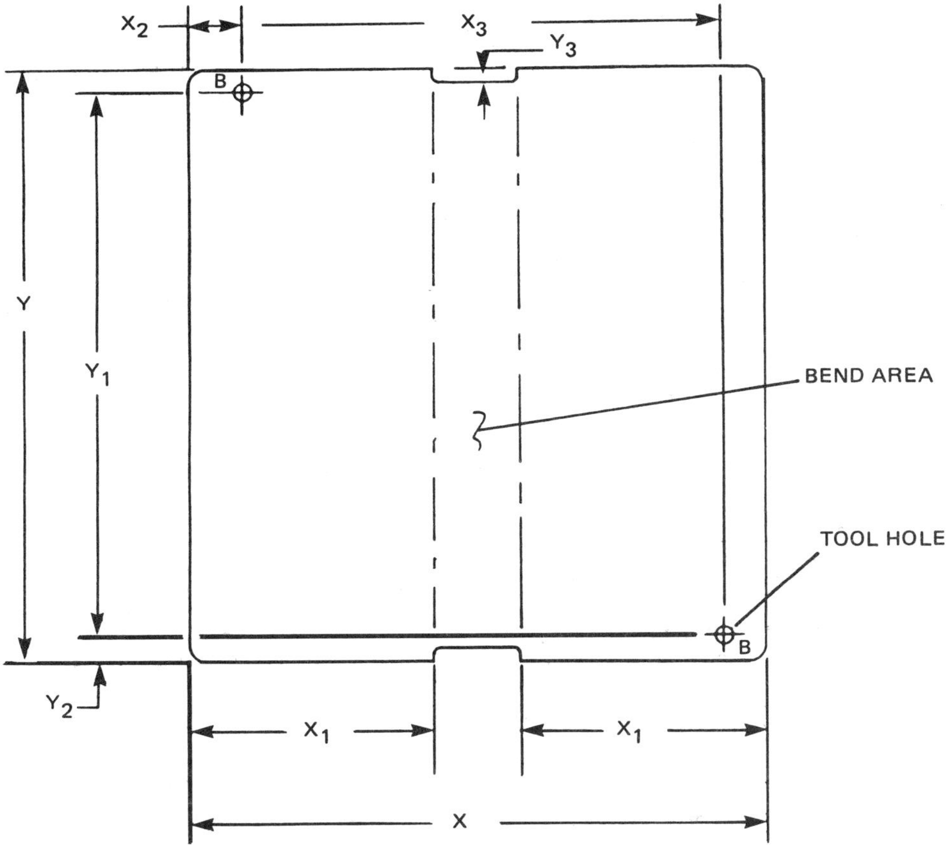

Fig. 6 Design layout showing internal tooling holes and the bend area.

erated art work masters and fabrication drawings which are very helpful in the manufacturing of a flex circuit substrate. Make sure to check your flex circuit supplier for CAD/CAM equipment compatability.

The only computerized equipment with software capability for designing flex circuits known to be useful to the design engineer is the CALMA Company (GE) system. This system offers the Design, Drafting, and Manufacturing Printed Circuits CAD system specifically for electrical and mechanical design. With this system, the mechanical layout task can be coordinated with electromechanical functions: schematic layouts, circuit verification, and flex circuit layout. By using a single database, the system can prevent design errors when creating electrical and mechanical documentation. The physical geometry of the assemble can be defined isometrically, and the flex circuit then integrated into the package. Since the exact locations of all the components are already in the CAD system database, information such as tooling hole dimensions, component sizes, and terminal patterns can be transferred directly to the flexible substrate assembly.

Location and orientation of SMDs

Another important consideration is the location and orientation of the SMDs on the flexible substrate. Most of the SMDs are supplied on a rectangular shape package by the majority of the vendors in sizes such as 50 mils by 80 mils and 60 mils by 120 mils for resistors and capacitors. The transistors, diodes, and some zener diodes are generally supplied in SOT-23 and modified SOT-23 packages (see Table 1).

Automatic placement of SMDs

It is important that some kind of geometrical pattern be established in order to minimize the head movement of the automatic SMD placement machine. Its accuracy can only be controlled by limiting the mounting head movement and its direction.

The best way to achieve this, and to take full advantage of the real estate, is to populate the flexible circuit layout area with the components in some type of symmetrical pattern, as in Fig. 7.

Notice that the SMDs are laid out so that the long side of the SMD faces the short side of the flexible substrate. This technique is recommended to minimize solder joint stress when the flex circuit is folded in that direction. Also, make sure that no devices are placed on the bendable areas to minimize any lifting pressure on the SMD solder pads.

If a rigid backer board is used in the flex construction, then the device's orientation can be rotated so that the longer side faces the longest axis of the flexible substrate, as in Fig. 8.

Unless the flex circuit design uses a rigid backer board, it is recommended that no SMDs be attached to the opposite side of the SMDs geometrical pattern. This will minimize any double-sided solder joint stress on the surface of the pad, since the thin Kapton polymide substrate would not be able to flex properly due to the restriction imposed by this assembly method.

Pad size and space requirements

Now, a few words about the pad size and the space requirement between leadless devices: The pad design for resistors and capacitors should be larger than the width of the leadless device. For example, if you are using a

Table 1 SMD Dimensions by Vendor

	Motorola		ROHM		Ferranti		Amperex		Nec	
Dimension	min	max	min	max	min	max	min	max	min	max
A	2.80	3.05	2.70	3.10		2.90	2.80	2.90	2.70	3.10
B	1.20	1.40	1.50	1.80		1.40	1.20	1.30	1.50[a]	
C	0.85	1.20	1.00	1.45		1.20	0.80	1.20	1.20	1.55
D	0.37	0.43	0.35	0.50	0.40[a]		0.37	0.43	0.35	0.50
F	0.076	0.127	0.09	0.25	0.10[a]		0.085	0.115	0.10	0.26
G	1.78	2.04	1.70	2.10	1.90[a]		1.90[a]		1.90[a]	
H	0.51	0.61								
K	0.10	0.25		0.15	0.05		0.10	0.20	0.10	0.25

L	2.10	2.50	2.60	3.00		2.75	2.10	2.50	2.20	3.00
M	0.45	0.61					0.40	0.60		
N	0.89	1.02	0.94	0.96	0.95[a]		0.95[a]		0.95[a]	

Source: ROHM Corporation

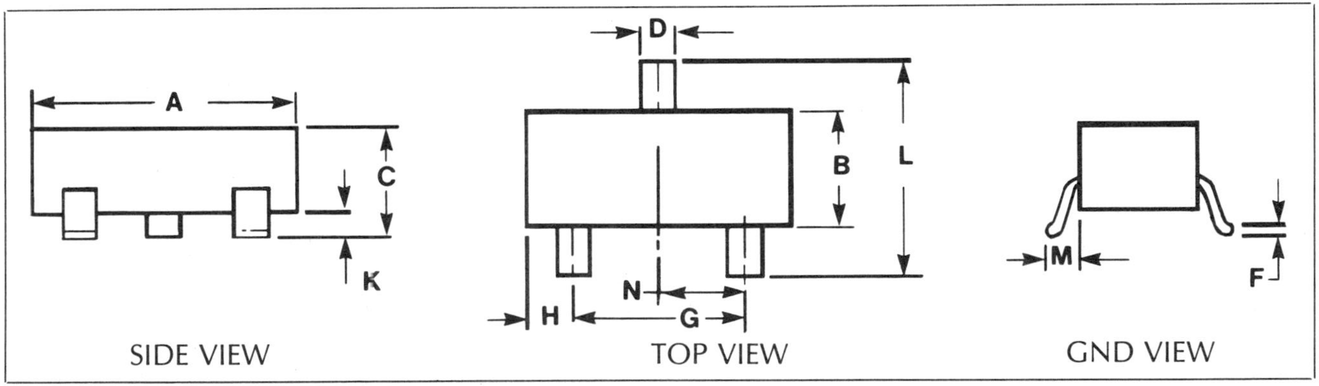

[a]Typical dimension

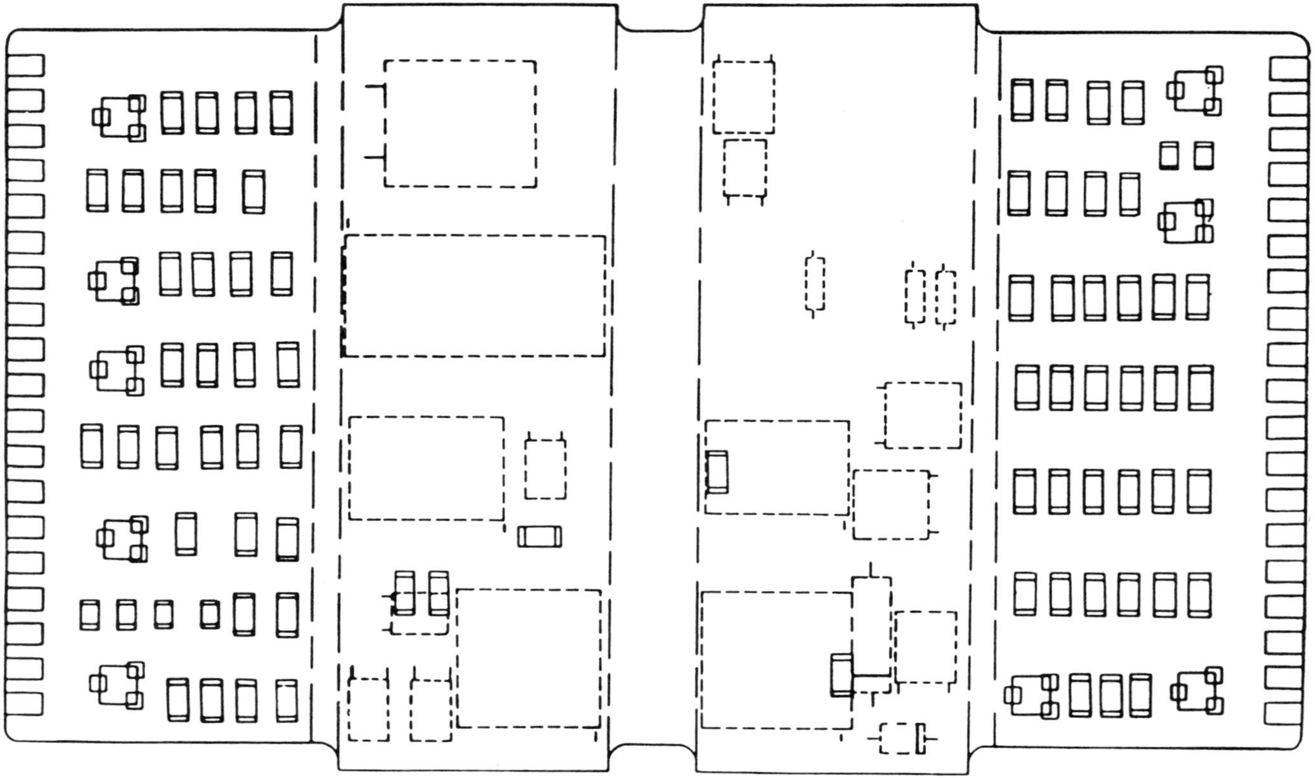

Fig. 7 Symmetrical layout of flexible circuit components.

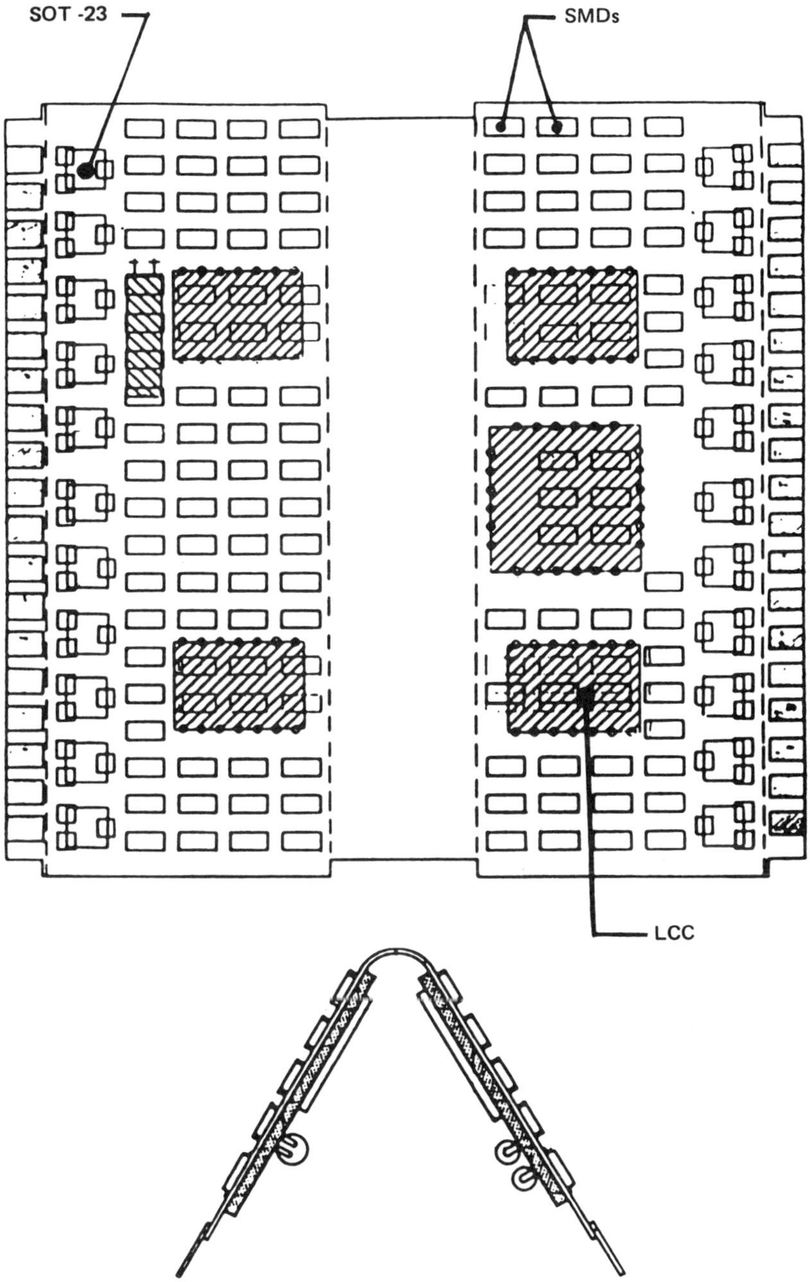

Fig. 8 Use of a rigid backer board allows the SMD longerside to face the longest axis of the flexible substrate.

50 mils by 80 mils leadless device, then the pad design should measure greater than 50 mils square. On the other hand, if the leadless component size is 60 mils by 120 mils, the pad should measure greater than 60 mils square.

The space between SMDs should be minimum of one-half the leadless device's width (Fig. 9). If the two adjacent components are different in sizes, the width of the larger part should be selected. For example, if the two leadless devices measure 50 mils by 80 mils and 60 mils by 120 mils, the space requirement will be 30 mils between the two components.

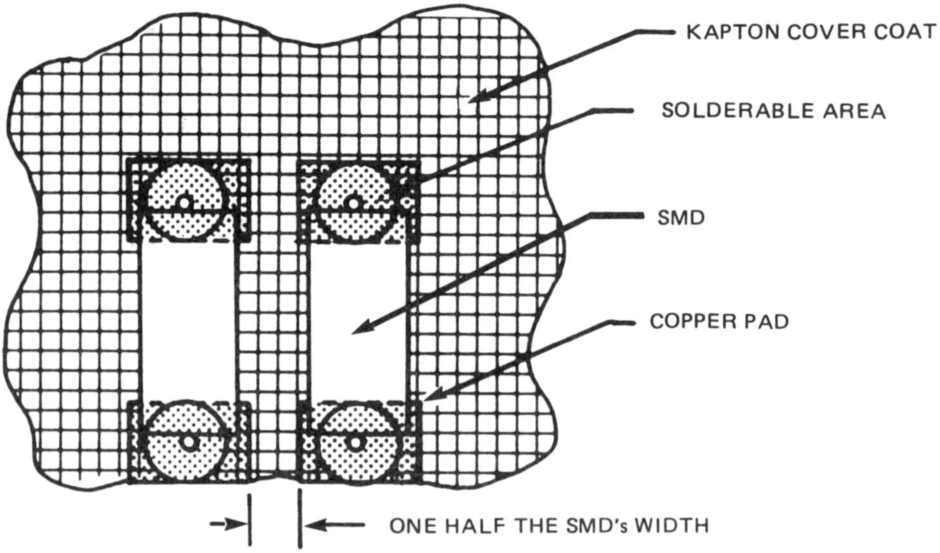

Fig. 9 The space between SMDs is one half of the device's width.

Placement of printed circuit traces

Another design consideration involves the placement of the printed circuit traces in the flexible substrate. It is suggested that the printed circuit traces in the selected side for affixing the SMDs be run parallel to the flexible substrate longest axis.

The far side of the flexible substrate should have printed circuit traces running in the direction parallel to the flexible substrate's shorter axis.

This placement technique of traces will not only provide easy circuitry flow, but it will also help prevent the I beam effect (see Fig. 10).

Leaded components

Let us now consider the use of leaded components. If any leaded device is required in the flex circuit design, the use of a rigid backer board is strongly recommended. This rigid backer material is used mainly to minimize component lead stress.

The rigid board material thickness will range between 15 and 30 mils, depending on your design criteria.

No printed circuit traces should be planned for this rigid backer board, since interfacing between the rigid backer board printed circuit traces and the flexible substrate printed circuit traces will require extra steps which will increase the rigid/flex assembly cost considerably.

The rigid backer board is generally laminated to selected areas of the flexible substrate and the apertures in the rigid backer board are usually larger than the holes found

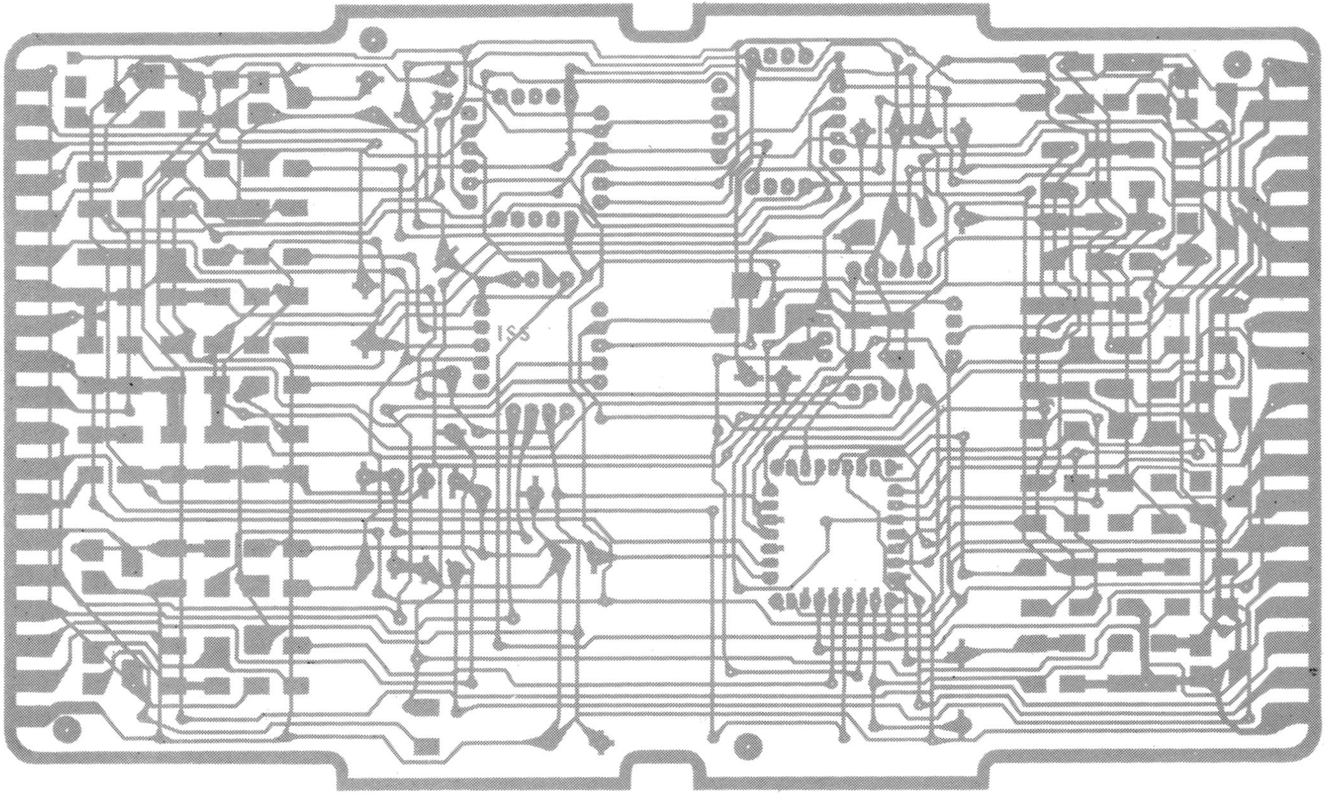

Fig. 10 Printed circuit traces technique. Copyright © Motorola, Inc.

in the flex base material. In some cases larger openings may be required to insure the proper affixing of parts on the surface of the flex circuit.

If a small leaded IC package is required, Kyocera Ltd. manufacturers a series of new dual-in-line packages with 50 mils lead-centers and in either 300 or 400 mils row spacing, depending on the number of pins required.

This new dual-in-line package can be wave soldered without any major problem. When planning the layout, it is recommended that an 8:1 scale be considered, so you can take advantage of using the 4:1 scale Rapid Design #320 drafting template for laying out the 50 mils leadcenters and the 300 mils dual-in-line row spacing.

Circuit schematic flow

Finally, we should discuss the circuit schematic flow. This is very important because when laying out a flex circuit with SMDs, the shorter the trace distance between the devices, the more real estate is available for the rest of the devices.

Make sure to group all the schematic functions according to design and identify all the circuits, especially if analog and digital electronic functions are found in the same flexible substrate. Another good idea is to use different series of reference numbers to locate and isolate the circuit functions. For example, use R101, C101, Q101, CR101 to identify parts of your regulator, and R301, C301, and U301 to identify the analog circuit.

3.2 Art Master Requirements for Leadless Devices

The flexible substrate art master requirements are similar to those of the printed wiring board. However, there are differences which will be covered in this chapter.

The scale selected for your flexible substrate art master can be either a 4:1, 8:1, or 10:1 scale. If your are planning to employ the use of a CAD system, chances are that a higher scale such as 20:1 might be used, depending on the size of the flexible substrate.

If a CAD system is employed to generate the art master, chances are that a new component library will be required since most of the CAD system libraries are stocked with hybrid component layout information.

In the hybrid technology, the pad outline is designed so that the component sinks into the solder paste pattern underneath the device. On the other hand, with the flexible substrate soldering method, the SMDs are soldered directly on the exposed device terminals (see Fig. 11).

Since the Flex-O-Pak design was manually laid out, the art master was also manually done. The art master was composed of the following five individual layers.

First art master layer

The first art master layer represents the flexible substrate solder side, which is the surface going over the dual wave solder. This layer houses all the discrete SMDs, such as the resistors, capacitors, and the SOT-23 packages used for diodes and transistors.

As mentioned earlier, the device's copper pads are laid out with a square shape. These

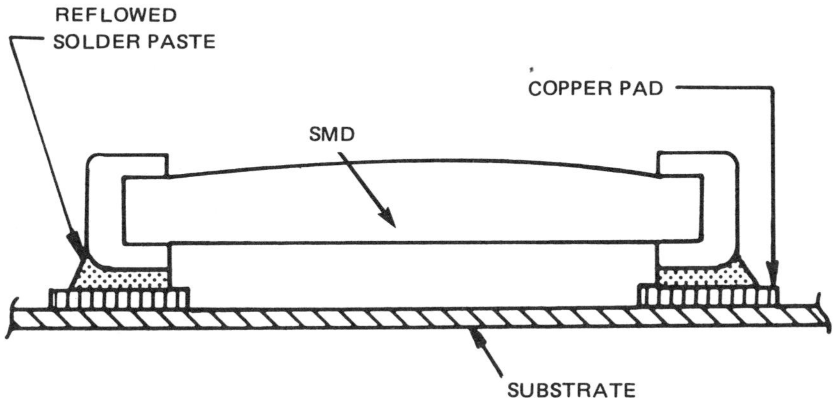

(a)

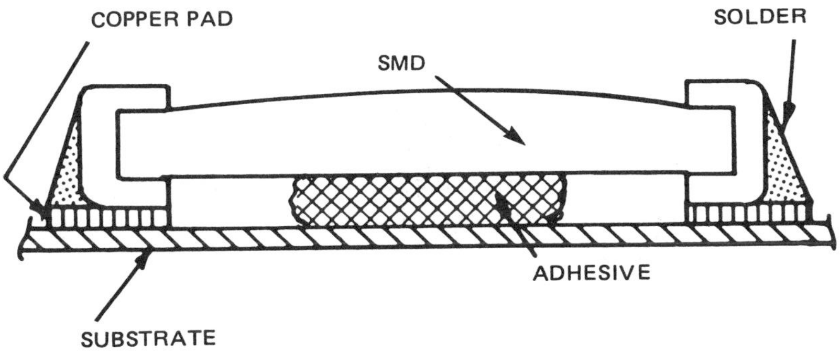

(b)

Fig. 11 (a) Hybrid technology; (b) flexible substrate.

square pads should have a center hole. The hole size will depend on the art master scale. This center hole has a dual role in the creation of the art master.

First, the hole is used for the visual art master alignment between the solder side and the corresponding cover coat side art master. The second function is that of a plated-through-via to the opposite side of the substrate if an interconnection is needed. Remember, if this technique is required, the plated-through-via must be covered with cover coat material on the opposite side of the flexible substrate. The finished plated-through-via should measure 16 mils in diameter. However, the hole diameter is not critical since it is only used for interconnection and not for a component lead.

The 1:1 art master should have a 25 mils grid increment to assure the right multiplier for the scale selected in your art master layout.

Since most of the SMD dimensions are in metric measures, it is recommended that you translate the measurement to the nearest hundredth of an inch.

If you select a scale which permits you to tape the art master with a 100 mils grid, then the flex circuit art master will be easy to tape manually.

The flexible substrate solder side layer has the bulk of the printed circuit traces. Because of this crowding, the average trace width should be about 10 mils with a minimum 10 mils space between the printed circuit traces.

It is suggested that the art master have 12 mils printed circuit traces with 8 mils spaces between the traces. Due to the photo reduction tolerances and the trace etching tolerances, the final printed circuit trace width will be between 1.5 and 2 mils less than the printed circuit trace width shown in the 1:1 art master (on a 1 ounce copper) (see Table 2).

The flexible substrate solder side art master layer should also include the datum tooling holes. A round pad area should be provided so that the flexible circuit vendor can punch the tooling holes for better accuracy than drilling.

If the need arises for the attachment of a lead frame or interconnecting pin, the flexible substrate solder side art master is the recommended layer. The recommended pad area for affixing the lead frame or pins should

Table 2 Effects of Etching, 0.6 EF

Line width before etching (mils)	Line width after etching (mils)	Percent reduction in line width
20	18.3	8.5
15	13.3	11.3
10	8.3	17.0
8	6.3	21.0
5	3.3	34.0

Etch factor (EF) defined as lateral etch in mils etched for each mil etched vertically. Figures based on clad removal per ounce.

Courtesy of Electronic Packaging and Production, June, 1984.

be proportional to the size and shape of the interconnecting device (see Fig. 12).

It is suggested that no printed circuit traces be laid out on the flexible substrate far side layer directly over the area where the lead frame or interconnect pin pads are located.

To determine the pad size and shape for the plated-through-vias, use the SMD square pad center hole for the plated-through-vias. If such a connection is not possible, the average plated-through-via pad should be as small as possible and the finished hole size should measure accordingly.

It is important to remember that all plated-through-vias are cover coated by the vendor, unless otherwise specified in the fabrication drawing.

When taping the printed circuit traces on the art master, makes sure the traces flow in an even way, and that all the minimum required spaces between the traces are observed, especially if they are traveling parallel for a long distance.

Do not bend the tape around the pad pattern or create any geometrical curvatures. This technique does not work well in flex circuit taping because the tape has a tendency to shift. It is better practice to use a 30 or a 45° angel. Do not use sharp bends or 90° corners. Figure 13 demonstrates the recommended method.

If an odd-pattern shape outline is required for the flexible substrate, make the outline border as heavy as possible for better detail. The flexible substrate solder side art master layer is the correct layer to show this outline border.

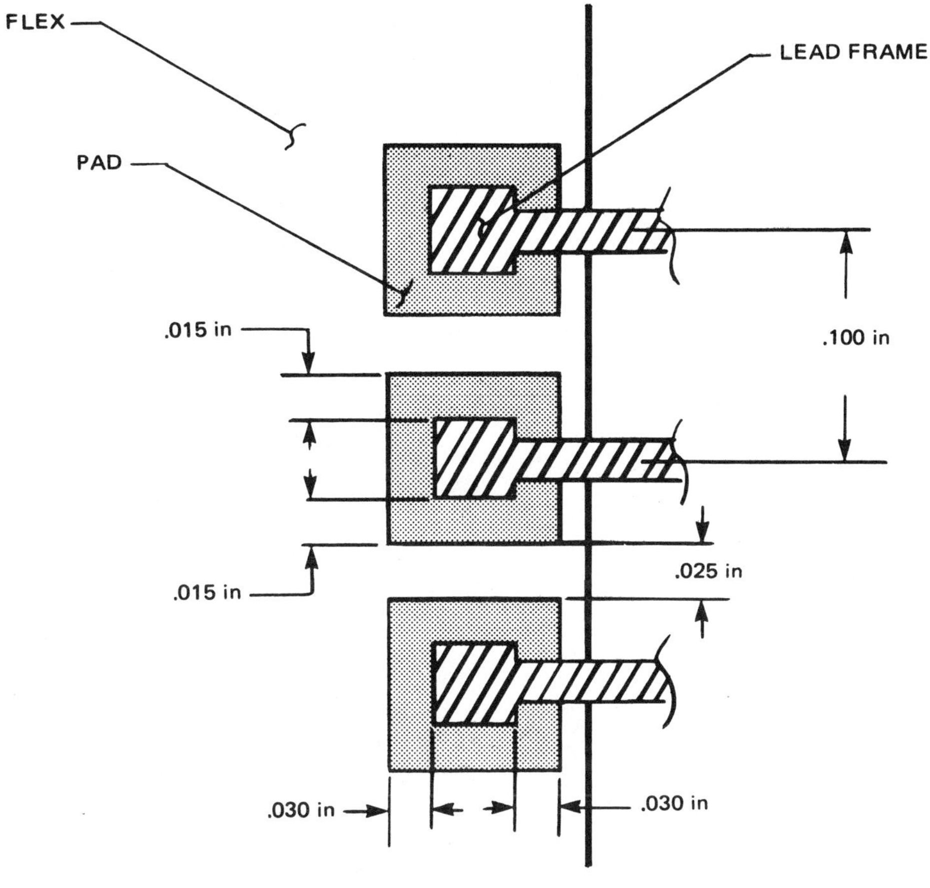

Fig. 12 Pad area proportional to the interconnecting device.

Leave a minimum of 50 mils between the edges of the traces or pad areas and the flex circuit edge. Make all the radii as generous as possible. The larger the radius, the less chance that curvature will be the weakest point in the design. Do not use square cor-

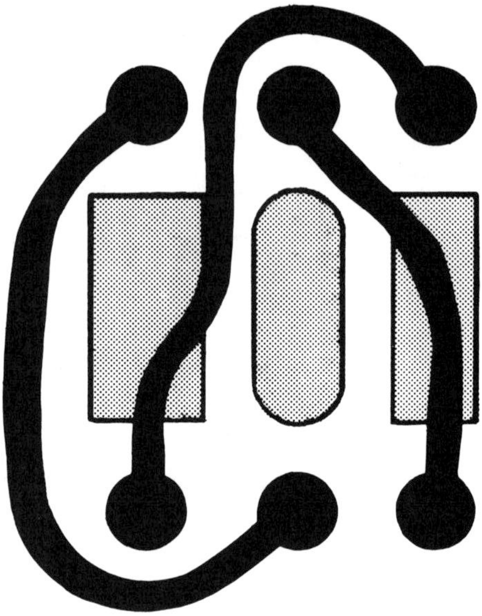

(a)

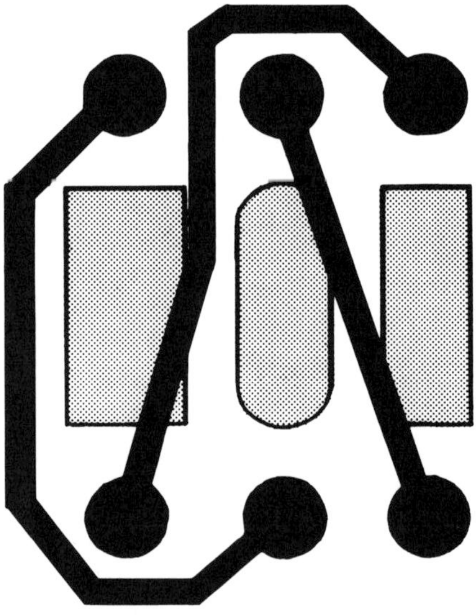

(b)

Fig. 13 Taping methods: (a) not recommended; (b) recommended.

ners or 90° angles, since this technique may add weaknesses to your flex circuit design. If there is a possible weakness in the design, and there is no possible way to avoid it, then the use of an inner glass-cloth reinforcement layer, laminated between the base material and the corresponding cover coat material is recommended.

Second art master layer

The second art master layer represents the solder side cover coat material. This art master layer will reflect the apertures that match the SMDs square pad centers in the flexible substrate solder side layer. This art master cover coat layer indicates the component copper pad areas that need to be exposed and free of any cover coat material.

Any pad located in the cover coat art master layer will provide an aperture in the cover coat material to create direct access to the flexible substrate surface. All exposed copper pads will be tin/lead plated by the flex circuit vendor.

The shape of the pads located in the cover coat art master layer will be round and the pad diameter will correspond to the diameter of the aperture required in the design on a 1:1 scale. For example, if the aperture desired should measure 50 mils in diameter, and your art master is being generated at a 4:1 scale, then the correct pad size should measure 200 mils in diameter. When selecting the diameter for the pad's inner hold size, try to make sure that the final aperture measures about 15 mils in diameter at a 1:1 scale.

The solder side cover coat art master layer will help the flex circuit vendor to determine if your aperture requirements will result in the solder surface area that you need in your design.

Fabrication drawing

Another document associated with these two art master layers is the farbrication drawing. Besides the art master layers, the fabrication drawing is the most important document because it shows the vendor what your finished flex circuit should look like and what restrictions the vendor is going to encounter throughout the construction of the flex circuit.

It is a good idea to create a composite of the two art master layers. The solder side art master layer should be photographed with a 40% halftone screen over a full contrast copy of the solder side cover coat art master. A photoprint of this composite should be incorporated into the fabrication drawing for better detail. Figure 14 is an example of such a photoprint.

It is also good practice to provide a hole legend in the fabrication drawing, detailing all information pertaining to the flexible substrate hole sizes as well as the cover coat material hole sizes for the SMD requirements. This chart will depict all openings whether they are for plated-through-vias or leaded components (see Table 3).

The fabrication drawing should also show all extra assembly operations for which the flex circuit vendor is responsible, such as pumicing of the flexible substrate surface, lead

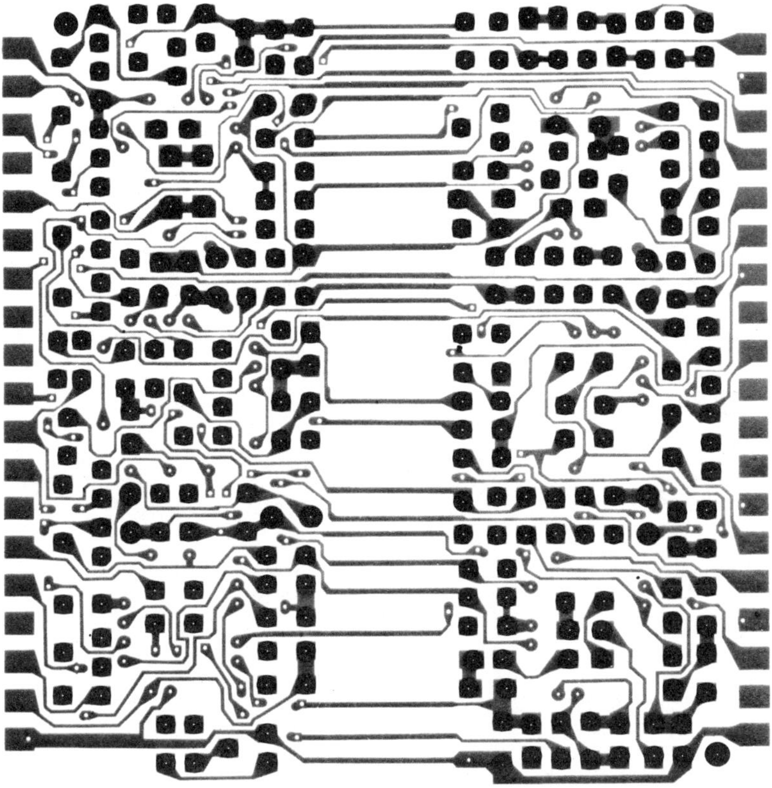

Fig. 14 Photoprint of composite of two art master layers. Copyright © Motorola, Inc.

frame attachment to the flex circuit, rigid backer board lamination, etc. It also details the material requirements, such as the flexible substrate base material (type and thickness), the cover coat material (type and thickness, if different from the base material), copper thickness, and the type of copper needed if

Table 3 Hole Legend

	Base laminate		Cover coat: through-holes		Cover coat: non-through-holes	
Symbol	Description	Quantity	Description	Quantity	Description	Quantity
Unmarked	.012 dia. max	275	—	—	.050 dia.	122
A	.025 dia.	12	.050 dia.	12	—	—
B	.062 dia.	2	.0635 dia.	2	—	—

sculptured fingers are used. The adhesive type should be identified in the fabrication drawing if different from the frequently used Pyralux adhesive.

Third art master layer

The third art master layer deals with the flexible substrate surface opposite the solder side. This layer is sometimes referred to as the component side layer. However, since there are components on both sides of the flexible substrate, it will be less confusing to call this layer the far side art master layer.

The far side layer will house all the leaded components as well as the LCC and any other devices that cannot be immersed in the molten solder. The number of traces on the far side surface are substantially less than on the flexible substrate solder side surface.

Like in the solder side layer, the pad geometry is the same and very seldom will it differ from the solder side pad shape. The inner holes of the pad provide alignment between the far side art master layer and the far side cover coat art master layer. They are also used for the front-to-back registration between the far side art master layer and the solder side art master layer.

If the Kyocera Ltd. dual-in-line IC package with the 50 mils lead centers is used, then the pad requirement should be taped as shown in Fig. 15.

If round pads instead of square pads are used in your layout make sure that the SMD pad diameter is larger than the aperture located in the cover coat art master layer. If the

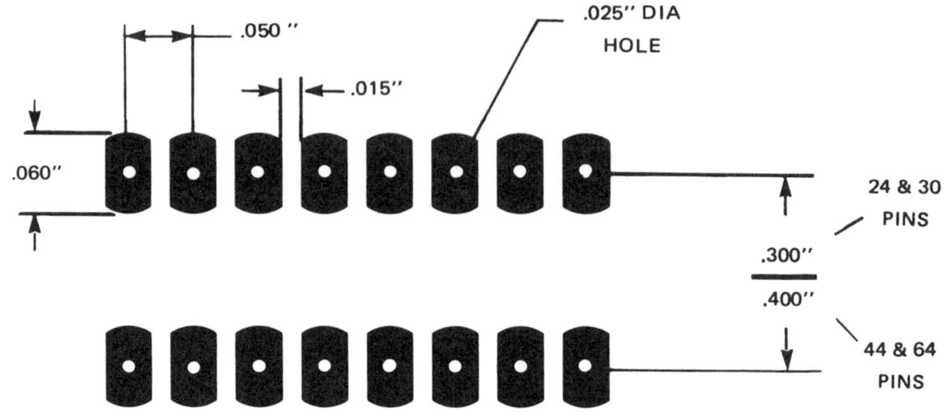

Fig. 15 Pad requirement for Kyocera Ltd. dual-in-line IC package.

square pad requirement cannot be observed, then the round pad should be anchored as securely as possible (see Fig. 16).

Fourth art master layer

The fourth art master layer is the far side cover coat layer. Like the solder side art master cover coat layer, the function of this layer is to provide apertures for the leaded components and the Chip carrier devices. The fabrication drawing information for this layer is the same as in the solder side layer.

Figure 17 shows a composite of the third and fourth art master layers which will help the flexible circuit vendor to fabricate the substrate with minimal problems.

Hole chart

A hole chart, similar to the one used in the solder side fabrication drawing, will be a

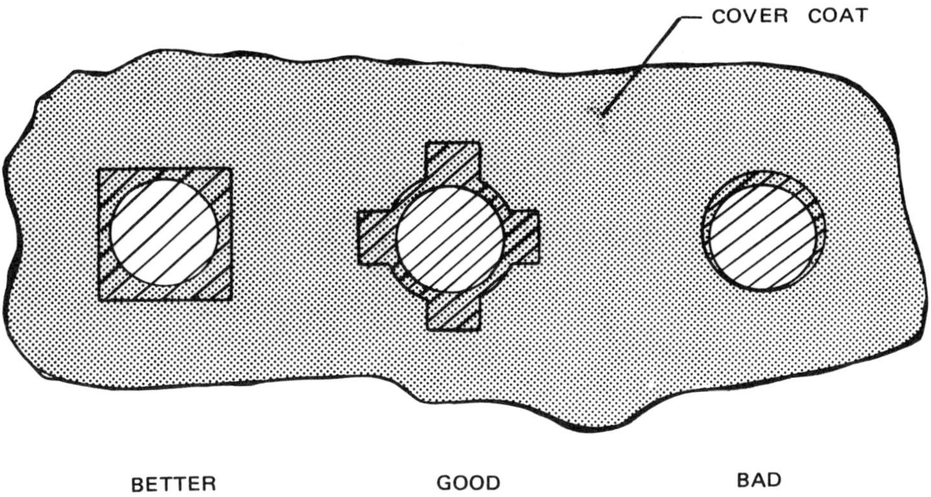

Fig. 16 Anchoring methods for round pads.

great help on the far side fabrication drawing because it is likely that you will require cover coat apertures different than those found on the solder side layer.

Fifth art master layer

The final art master layer details the component assemblies for both sides of the flexible substrate. This art master layer is not mandatory, but it will help the flexible circuit vendor to understand the location of the SMDs as well as the leaded components.

If you follow these recommendations when creating your art masters and the associated fabrication drawings, the result of your design work will be clearly understood and your flexible circuit vendor will be able

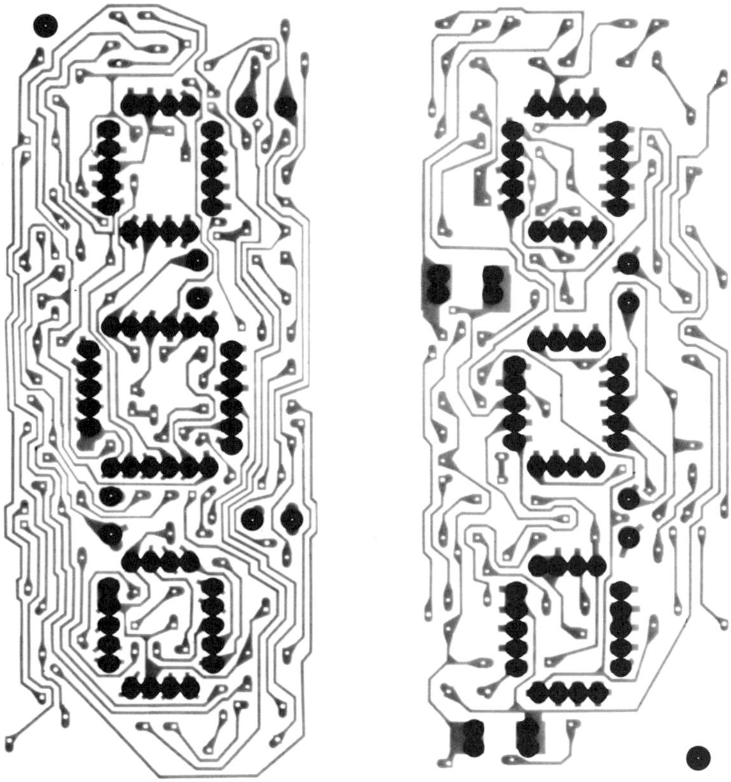

Fig. 17 Composite of third and fourth art master layers. Copyright © Motorola, Inc.

to help you better because he will be fully aware of your design needs and criteria.

The information reviewed in this chapter is the result of seven years of working with flex circuits, and three years of developing new assembly techniques to use SMDs on flexible substrates. The compiled data has been proven in practice, and it is hoped that this chapter has provided food for thought to help you develop even better assembly techniques in the near future.

4
PROCESSING TECHNIQUES

4.1 Soldering of SMDs on Flexible Circuits

Surface Mount Devices (SMDs) have already started to change most aspects of electronic assembly. Unlike the mature processes used to produce conventional printed wiring board assemblies, surface mount technology is not yet well developed.

Surface mount technology

In this new technology, a component is soldered to the solder pads of a substrate that match the footprints made by the leads of the component package. Surface mounting has several advantages over the method that it replaces. For instance, you can easily double substrate density by using smaller components and eliminating most plated-through-holes.

Surface mounting also allows components to be placed on both sides of a printed circuit substrate—a major advantage. In fact, the use of chip capacitors and resistors, plastic leaded chip carriers, and other devices could in theory give these substrates packaging densities equal to, or better than, those of hybrids.

Finally, surface mounting offers still another advantage: better performance. The mixture of passive chip devices and tiny leaded active components reduces the length of the leads between interconnections, radically lowering the parasitic capacitance and inductance of the printed circuit substrate wiring.

One key reason for the growth of surface-mounting is the availability of surface-mountable active and passive components. Over the last few years, a low profile revolution in IC packaging has supplied the fundamental components for surface mounting/

miniaturized IC and semiconductor packages.

Surface mounting, however, is based on ICs and discrete semiconductors with leads on 50 mils centers and tiny chip components. This requires finer lines and spaces and small vias rather than through-holes.

This is the reason why the use of flex circuit substrates becomes very appealing to the design engineer. By using Kapton polymide film for the flex circuit substrate, we retain the properties and benefits of a conventional polyimide printed circuit board while adding flexibility. The reduced thickness of the Kapton substrate makes soldering of surface mount devices easier and less expensive.

The two principal technologies in use today for surface mount attachment are the solder paste/reflow method and the adhesive attach/wave solder approach.

The solder paste/ reflow method

The solder paste/reflow method is used with traditional hybrid circuit techniques and places SMDs on the top of the substrate. The first step is to put the solder paste onto the substrate. The first step is to put the solder paste onto the substrate using either a screen print or a stencil process. Components are then placed by machine into the screened paste, which is then dried to drive off carrier solvents. The solder paste is then reflowed by one of several common methods, including infrared (IR), vapor phase, or thermal conduction heating. Vapor phase is the only safe process for the substrate recommended by DuPont, the manufacturer of Kapton, since the other two processes will expose the flex circuit material to extensive high heat, which could harm its integrity. The flexible substrate then goes through a cleaning step to remove residual flux and any stray solder balls that may be on the substrate.

If leaded components are needed, they are inserted by hand. The flexible substrate assembly can be soldered by hand or conventional wave solder at a temperature lower than the melting point of the solder paste first used.

The adhesive attach/wave solder method

The adhesive attach/wave solder approach involves affixing SMDs to the underside of the flexible substrate using special adhesives. The adhesive used to hold the SMDs must meet certain basic requirements. Thermoset structural adhesives such as epoxies or acrylic compounds are normally used. Curing can be performed by application of either heat or ultraviolet energy (UV). A flexible circuit assembly must be baked prior to soldering as recommended by the laminate manufacturer to drive off residual moisture.

The adhesive can be applied using one of several techniques. Most commonly it is applied by a nozzle on the pick-and-place machine immediately prior to placing the SMDs. To eliminate the need for time consuming adjustments to the dispensing system, a single drop size should be used for all applications whenever possible. The resistors and capacitors have a nominal area for adhesive application, while on the SOT-23 the adhesive has to fill a bigger gap between the flexible substrate surface and the bottom of the device.

Once the components are affixed, they are ready to be wave soldered. Wave soldering of SMDs assemblies, however, is not identical to the process used for leaded component assemblies. The standard waves used to solder leaded components are not effective when used for SMDs. The wave solder manufacturers have addressed this issue and now offer various modifications to their standard equipment.

The use of a dual-wave that works in sequence overcomes the problems inherent in conventional wave soldering techniques when applied to leadless devices.

The first wave insures that the hot molten solder, if forced into the leadless devices, reaches all attachment points. The narrow contact length permits gas evolved from either the flux or the adhesive to escape.

The secondary wave provides the needed controllable exit conditions for the flexible circuit substrate. Its flow is in the same direction and at the same relative velocity as the flexible substrate assembly travel. This minimizes turbulence caused by the leadless devices.

Preheating is a critical part of the SMDs soldering process. In addition to eliminating flux generated gases, the flexible substrate must be adequately preheated so that the conveyor speed will limit the dwell time in the solder wave to no more than three to four seconds. This maximum dwell time is necessary to prevent loss of connecting metallization on the leadless components.

Ideally, the preheat should bring the temperature of the flexible substrate top side to between 200 and 220°F (93.5–104.5°C) just prior to entering the solder wave.

Using the dual-wave soldering system, SMDs can readily be combined with leaded components on the same flexible substrate. One of the flex circuit design guidelines on mounting leadless devices mentioned in Chap. 3 is to separate the leaded components from the leadless parts. Normally leadless devices are affixed after the placing of the leaded components. This sequence insures that the SMDs are not damaged by the cut-and-clinch tool if an automatic inserter is employed (see Fig. 1).

Cleaning

Cleaning is normally one of the final assembly processes; therefore, it is one of the last

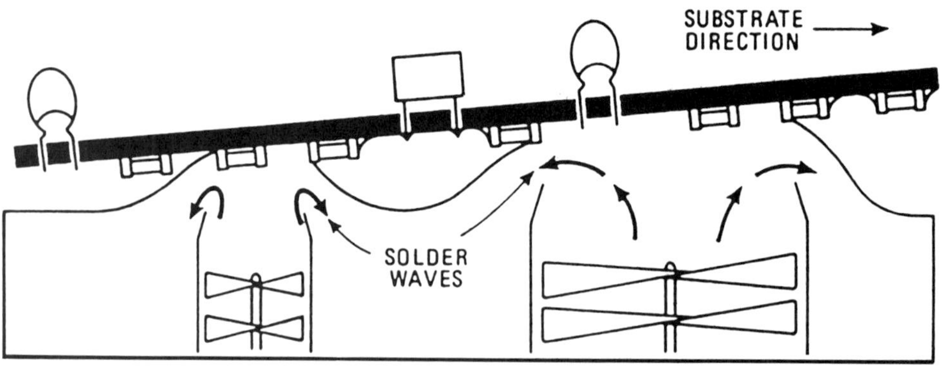

Fig. 1 Dual-wave soldering system. Courtesy of *Electronics*, February, 1984.

steps which can affect the quality and reliability of the circuit.

The cleanliness of the flexible substrate assembly, both around and underneath the components, is essential. In cleaning SMDs, the most important task is removal of fluxing agents and flux carriers after the solder process. The difficulty of removing the flux residues from SMDs increases dramatically as the space underneath the leadless device decreases, and as the leadless device density, device surface area, and flux activation increase (see Table 1).

The flexible substrate design and the component layout can facilitate the task of cleaning the SMDs. Cleaning difficulties and field failures can result if the flexible substrate layout does not adequately address these potential problems, as covered in Chap. 3.

Repairing

Repair of surface mount assemblies is another often neglected aspect of the new SMD technology. Unlike leaded components, SMDs are not readily removed with a simple soldering iron and desoldering tool. This is especially true for leadless devices which have been affixed using the adhesive approach.

On the flexible substrate, overheating can cause lifting of the pad metal. However, with the proper penciltip soldering iron, the removal of SMDs works quite well for small components.

Attaching a SMD replacement is more complicated. Particularly with multileaded LCC, it is difficult to align all the pins with

Table 1 Cleaning Difficulty of Some SMDs

SMD type	Typical component varieties	Component types	Assembly cleanliness difficulty
Type I	Passive-discrete Active-discrete	Chip resistors and capacitors, SOT-23, SOT-89	Easiest
Type II	Integrated circuit packages (leaded)	Flat packs SOIC Quad-Packs Leaded chip carriers	Difficult
Type III	Integrated circuit packages (leadless)	Leadless chip carriers	Very difficult
Type IV	Passive-discrete Active-discrete Integrated circuit packages Conventional devices	All types	Very difficult

Source: Courtesy of *Electronic Packaging and Production*, August, 1984.

the metal pads. However, with the new LCC soldering technique (which will be discussed in Section 4.3), the same soldering process can be used to repair and replace leadless devices.

Although surface mount technology is gaining momentum rapidly, several industry areas, such as flex circuit manufacturing, need considerable attention and new developments in order to support all the new innovative designs which are sure to be developed in the near future.

4.2 Affixing of Interconnection Pins on Flex Circuits

As mentioned in Chap. 2, the interconnecting of flexible substrates has been a major problem faced by many design engineers.

There are several methods used in the interfacing of a flexible substrate to a main rigid board or chassis assembly. However, until recently only two basic techniques for interfacing were available to the design engineer.

Jumper flex circuits

This method is widely used in the interfacing of two rigid boards. The jumper flex circuit is quickly installed, and may be wave soldered or plugged into several types of sockets, with limited insertion cycling. However, most of the time these flex circuit assemblies are component-free and their design layout is mainly single-sided, due to the manufacturing technique employed.

Wide printed wiring traces

The second method used to interface a flex circuit assembly with a rigid printed wiring board is designing wide printed writing traces on the flex circuit surface. The edge of the flex circuit is then left without covercoat material for a predetermined distance and the exposed copper is plated. This particular method for terminating flex circuits lends itself to several readily available commercial connectors.

Multiple-crimp technique

Now that the mechanical and electrical quality of the multiple-crimp technique has been tested, a wide range of applications has become possible.

Multiple-crimp contacts rely to some degree on accepted and proven technology, especially as far as materials and plating are concerned. However, this technique is recommended and applicable mostly to single-sided flex circuit layouts.

Flex-O-Pak pins technique

Since none of the innerfacing methods available were compatible with the Flex-O-Pak substrate, the interconnecting of the optional module was accomplished with the cooperation of the flex circuit manufacturer Parlex Corporation Inc., Methuen, Mass.

We began to explore several materials which were readily available and could lend themselves to welding or high temperature soldering to soft annealed copper pads used in the flexible substrate (Table 2).

To achieve the needed reliability, the solder alloy must possess corrosion resistance, strength, ductility, and resistance to fatigue. Also, the alloy needs to have a comparable thermal coefficient of the soldered materials. This helps to minimize the effects of thermal stresses.

The selection of the solder alloy is also dependent on the method of soldering. Such factors as melting point, plastic range, speed of wetting, ability to wet or spread and bond to different metals, and appearance after soldering must be considered when selecting the solder alloy (see Fig. 2). The designer must be aware that selection of the solder alloy is only part of the solder system of interdependent materials and properties, including fluxes, surfaces to be soldered, soldering method, and postsoldering cleaning techniques.

The interconnecting lead frame material was narrowed down to a copper alloy. The leads were coated with a nickel/gold plating process.

Since this flexible substrate was to be plugged into sockets in the main radio board, some sort of additional support was required to prevent pin breakage as well as to maintain pin integrity as the module is inserted into the main radio board. It was decided that support header assembly was required.

Another objective of this flexible substrate was to be repairable. The header assembly had to be detachable from the flexible substrate assembly, unlike the method employed in affixing hybrid headers.

To prevent the pin solder joints from reflowing as the flexible substrate assembly ran over the wave solder pot, a higher temperature solder than the one found in the solder pot (485°F/253°C) was needed.

In order to prevent this problem, the lead frame is affixed to the flex circuit pad surface using a high temperature solder paste.

Since the Flex-O-Pak substrate has thirty-eight connections, nineteen leads on 100 mils centers at each end of the flexible substrate, it was determined by Parlex Corp. that an automatic solder system was needed to handle this high temperature process, speed production, and achieve a reasonable labor piece price.

Parlex Corporation has developed a high temperature soldering system capable of af-

Table 2 Solder Compositions, Percent

Alloy grade	Tin desired	Lead, nominal	Antimony			Silver			Bismuth, max	Copper, max	Iron, max	Aluminum, max	Zinc, max	Arsenic, max
			Min	Desired	Max	Min	Desired	Max						
70A	70	30	—	—	0.12	—	—	—	0.25	0.08	0.02	0.005	0.005	0.03
70B	70	30	0.20	—	0.50	—	—	—	0.25	0.08	0.02	0.005	0.005	0.03
63A	63	37	—	—	0.12	—	—	—	0.25	0.08	0.02	0.005	0.005	0.03
63B	63	37	0.20	—	0.50	—	—	—	0.25	0.08	0.02	0.005	0.005	0.03
60A	60	40	—	—	0.12	—	—	—	0.25	0.08	0.02	0.005	0.005	0.03
60B	60	40	0.20	—	0.50	—	—	—	0.25	0.08	0.02	0.005	0.005	0.03
50A	50	50	—	—	0.12	—	—	—	0.25	0.08	0.02	0.005	0.005	0.025
50B	50	50	0.20	—	0.50	—	—	—	0.25	0.08	0.02	0.005	0.005	0.025
45A	45	55	—	—	0.12	—	—	—	0.25	0.08	0.02	0.005	0.005	0.025
45B	45	55	0.20	—	0.50	—	—	—	0.25	0.08	0.02	0.005	0.005	0.025
40A	40	60	—	—	0.12	—	—	—	0.25	0.08	0.02	0.005	0.005	0.02

40B	40	60	0.20	—	0.50	—	—	—	0.25	0.08	0.02	0.005	0.005	0.02
40C	40	58	1.8	2.0	2.4	—	—	—	0.25	0.08	0.02	0.005	0.005	0.02
35A	35	65	—	—	0.25	—	—	—	0.25	0.08	0.02	0.005	0.005	0.02
35B	35	65	0.20	—	0.50	—	—	—	0.25	0.08	0.02	0.005	0.005	0.02
35C	35	63.2	1.6	1.8	2.0	—	—	—	0.25	0.08	0.02	0.005	0.005	0.02
30A	30	70	—	—	0.25	—	—	—	0.25	0.08	0.02	0.005	0.005	0.02
30B	30	70	0.20	—	0.50	—	—	—	0.25	0.08	0.02	0.005	0.005	0.02
30C	30	68.4	1.4	1.6	1.8	—	—	—	0.25	0.08	0.02	0.005	0.005	0.02
25A	25	75	—	—	0.25	—	—	—	0.25	0.08	0.02	0.005	0.005	0.02
25B	25	75	0.20	—	0.50	—	—	—	0.25	0.08	0.02	0.005	0.005	0.02
25C	25	73.7	1.1	1.3	1.5	—	—	—	0.25	0.08	0.02	0.005	0.005	0.02
20B	20	80	0.20	—	0.50	—	—	—	0.25	0.08	0.02	0.005	0.005	0.02
20C	20	79	0.8	1.0	1.2	—	—	—	0.25	0.08	0.02	0.005	0.005	0.02
15B	15	85	0.20	—	0.50	—	—	—	0.25	0.08	0.02	0.005	0.005	0.02
10B	10	90	0.20	—	0.50	—	—	—	0.25	0.08	0.02	0.005	0.005	0.02
5A	5[d]	95	—	—	0.12	—	—	—	0.25	0.08	0.02	0.005	0.005	0.02
5B	5[d]	95	0.20	—	0.50	—	—	—	0.25	0.08	0.02	0.005	0.005	0.02

(continued)

Table 2 *(continued)*

Alloy grade	Tin desired	Lead, nominal	Antimony			Silver			Bismuth, max	Copper, max	Iron, max	Aluminum, max	Zinc, max	Arsenic, max
			Min	Desired	Max	Min	Desired	Max						
2A	2[e]	98	—	—	0.12	—	—	—	0.25	0.08	0.02	0.005	0.005	0.02
2B	2[c]	98	0.20	—	0.50	—	—	—	0.25	0.08	0.02	0.005	0.005	0.02
2.5S	0[f]	97.5	—	—	0.40	2.3	2.5	2.7	0.25	0.08	0.02	0.005	0.005	0.02
1.5S	1[g]	97.5	—	—	0.40	1.3	1.5	1.7	0.25	0.08	0.02	0.005	0.005	0.02
95TA	95	0.20 max	4.5	5.0	5.5	—	—	—	0.15	0.08	0.04	0.005	0.005	0.05
96.5TS	96.5	0.20 max	0.20	—	0.50	3.3	3.5	3.7	0.15	0.08	0.02	0.005	0.005	0.05

[a]Analysis shall regularly be made only for the elements specifically mentioned in the above table.

[b]The chemical requirements of SAE specifications Nos. 1A, 2A, 2B, 3A, 3B, 4A, 4B, 5A, 5B, 6A, and E-07 conform substantially to the requirements for alloy grade Nos. 45B, 40B, 40C, 30B, 30C, 25B, 25C, 20B, 20C, 15B, and 2.5S, respectively.

[c]Federal specifications are similar to the above alloy grade Nos. 70B, 63B, 60B, 50B, 40B, 35C, 30C, 20C, 2.5S, 1.5S, and 95TA.

[d]Permissible tin range, 4.5 to 5.5 percent.

[e]Permissible tin range, 1.5 to 2.5 percent.

[f]Tin maximum, 0.25 percent.

[g]Permissible tin range, 0.75 to 1.25 percent.

Note: Solder Metal Specification ASTM B-32 is printed in full in the latest *Annual Book of Standards Part 8*, American Society for Testing and Materials, 1916 Race Street, Philadelphia, Pennsylvania 19103.

Source: Courtesy of *Electronic Packaging and Production*, July, 1979.

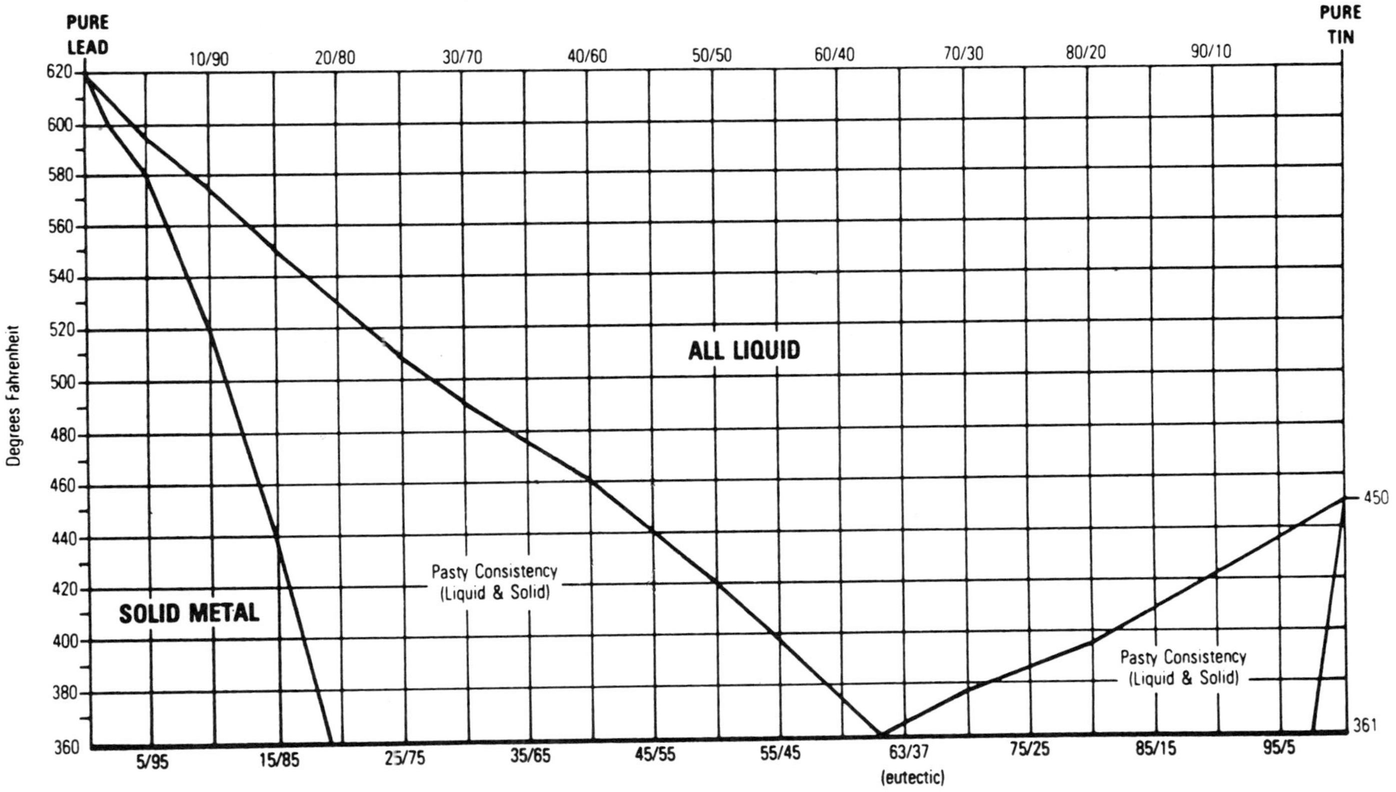

Fig. 2 Solder alloy metal composition showing melting point. Courtesy of *Electronic Packaging and Production*, July, 1979.

fixing the thirty-eight lead connections in about two minutes time. This automatic, high temperature solder equipment not only makes the solder process reliable, but also makes it affordable.

Another concern was the retention of the flex circuit lead assembly within the header apertures. This problem was solved by using an aggregate height above the location of the leads with layers of acrylic adhesive and thick Kapton cover coat material strip. The cover coat material strip covers about 100 mils from the flexible substrate's edge and runs across the long axis of the flexible substrate (see Fig. 3).

The total thickness needed over the interconnecting pins was about 30 mils minimum. This allowed the Flex-O-Pak pins to interlock inside the expandable apertures in the header design. Figure 4 shows a cross-section of this interlocking ability.

The flexible substrate assembly is inserted into the plastic expandable header apertures, which are arranged in two rows of nineteen pins each. The apertures are capable of accepting and retaining, with the help of the flexible substrate edge cover coat material strip buildup, each of the thirty-eight interconnecting pins.

Since the pin soldering process was an innovation in the design of the flexible substrate, it was the key in making the flexible substrate a pluggable module assembly. This has also opened the door to new ways of interfacing flexible substrates with rigid board assemblies.

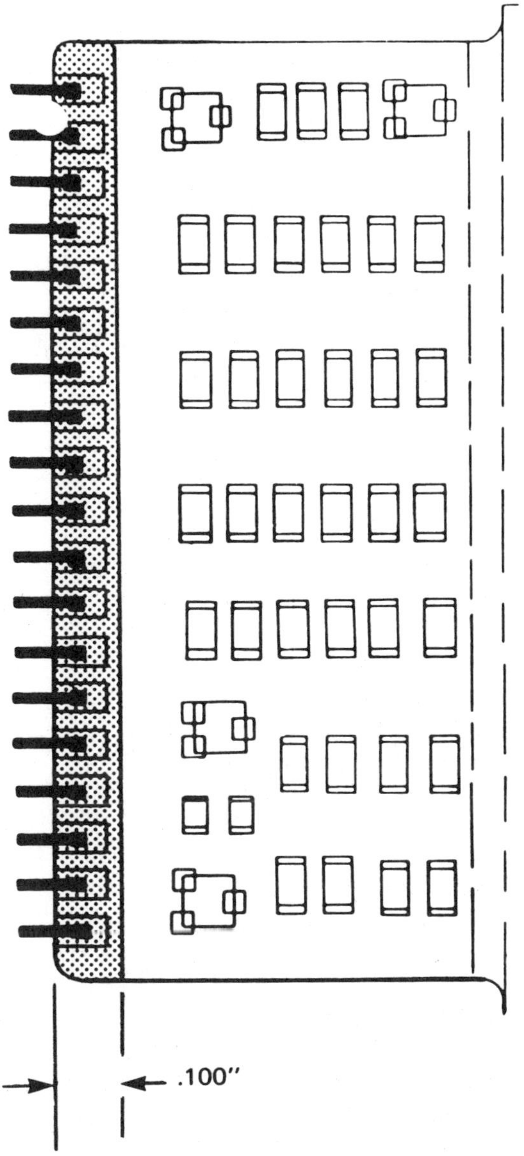

Fig. 3 Cover coat material strip.

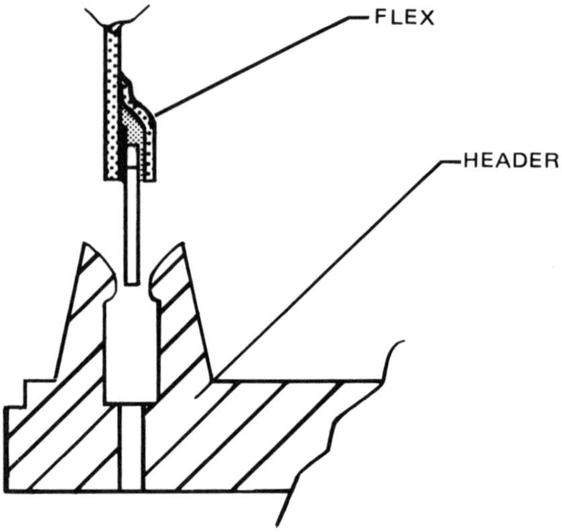

Fig. 4 Interconnecting pins in the header design.

4.3 Chip Carrier Affixing and Soldering on Flex Circuits

This section deals primarily with the steps required to affix an integrated circuit chip carrier on the top side of a flexible printed circuit substrate and the capillary technique used to solder this device on a dual wave solder system.

The present industry state-of-the-art technology for soldering chip carriers on a rigid substrate relies on screening or stenciling the performed solder paste on a layout geometrical pattern and then reflowing the substrate assembly with an IR or a vapor phase system. The technical process must be repeated if SMD on a double-sided substrate are being used.

Using the capillary soldering technique

The capillary technique allows the affixing of the chip carrier on the top side of the flexible substrate, and, at the same time, wave soldering of other, leadless devices on the bottom side minimizing the time exposure to high heat and the problems created by coefficient mismatch of the different devices and the flexible substrate.

In the past decade, the electronic industry has evolved from hybrids and leaded component printed circuit boards to a mixture of both technologies.

Because most components have been miniaturized and SMDs generally employ less metal and epoxy then their leaded counterparts, their ability to withstand mechanical stress is limited.

The reasons behind the growing popularity of SMDs are the cost and space savings, or more likely, a combination of the two. Although surface mounted devices contribute to the space savings made possible by the surface mounted technology, it is the reduction in the integrated circuit size that pays the greatest dividends.

It is because of these technological changes that a new, less expensive method to solder the chip carrier was needed. The proposed capillary soldering technique will help the electronic industry to assemble and to solder the chip carrier without the use of expensive IR or vapor phase equipment.

At first, chip carrier soldering was limited to ceramic hybrids. In the past five years, the electronic industry has been experimenting with polyimide printed wiring boards as a potential material replacement for the ceramic hybrids.

The substrates used by the electronic industry have experienced a solder joint reliability problem. The thermal coefficient of expansion mismatch between the ceramic chip carriers and the printed wiring boards creates stress problems and results in low reliability and yields.

The flexible material familiar to the electronic industry is Kapton polyimide film type H with a thickness range between 1 and 5 mils.

This flexible polyimide material and soft rolled, annealled copper conductor have opened up new design opportunities to the design engineer.

With the proper base material thickness of Kapton, such as 1 mil, and one ounce of rolled, annealled copper; a double-sided flex circuit can be designed and built.

Besides saving space, this flexible substrate offers the design engineer the flexibility needed to handle passive leadless devices, as well as chip carriers on a dual wave soldering system.

When we combine this flexible substrate with the small SMDs (typically 50 mils by 80 mils), we have achieved a design freedom never before available to the industry. This flexible substrate has memory and minimum TCE mismatch with the leadless devices, and it is thin enough to allow the use of a capillary soldering technique.

Let us take a basic electronic circuit design with several chip carriers and other passive leadless devices. The chip carriers are rectangular in shape with about eighteen leadless connections. The center of the pad is 50 mils. No corner pad connections are recommended, and the total chip carrier area is less than ½ inch square. Most of the remaining components are SMDs, such as resistors, capacitors in 50 mils by 80 mils packages, transistors, and diodes on a modified SOT-23 package. Because of some value restrictions, we are forced to use leaded components, such as electrolytic capacitors, and a small crystal.

Due to the design restrictions, most design engineers would use some kind of printed circuit hard substrate and, with no other choice, an IR reflow or vapor phase soldering system. Additional steps would be needed to manually insert the leaded components, adding cost.

With the proper fixturing and the capillary soldering technique, this packaging design can be run over a conventional dual wave solder system.

Figure 5 shows a flex circuit assembly. The solder side of the flexible substrate contains all the passive SMDs laid out geometrically as mentioned in Chap. 3. On the far side of the same flexible substrate you can see 6 the chip carriers' geometrical footprints as well as the other leaded components.

The most important advantage of this flex circuit layout is the automatic placement of most of the SMDs. Since the flexible substrate is small, it is most efficient to assemble the circuit patterns so that a number of flexi-

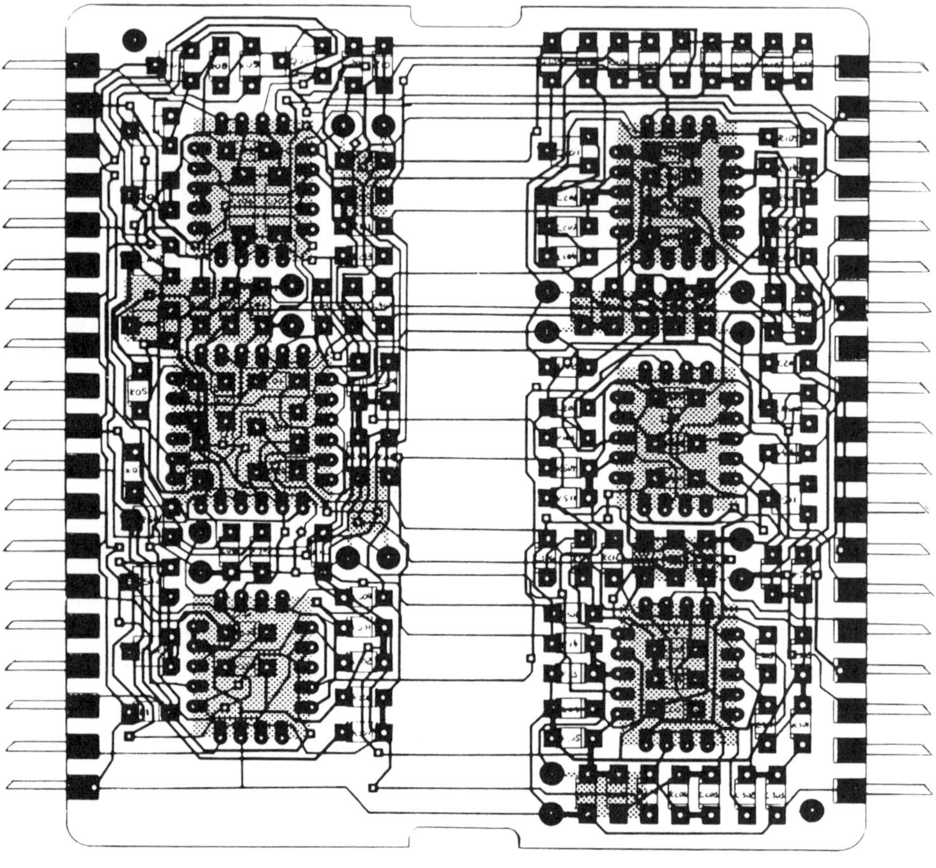

Fig. 5 View from solder side. Copyright © Motorola, Inc.

ble substrates can be processed over the dual-wave soldering system at the same time (see Fig. 6).

On this flexible substrate design, four flex circuits were selected and a rigid board carrier was used. The rigid board carrier material is FR-4 glass epoxy non-copper clad, both sides.

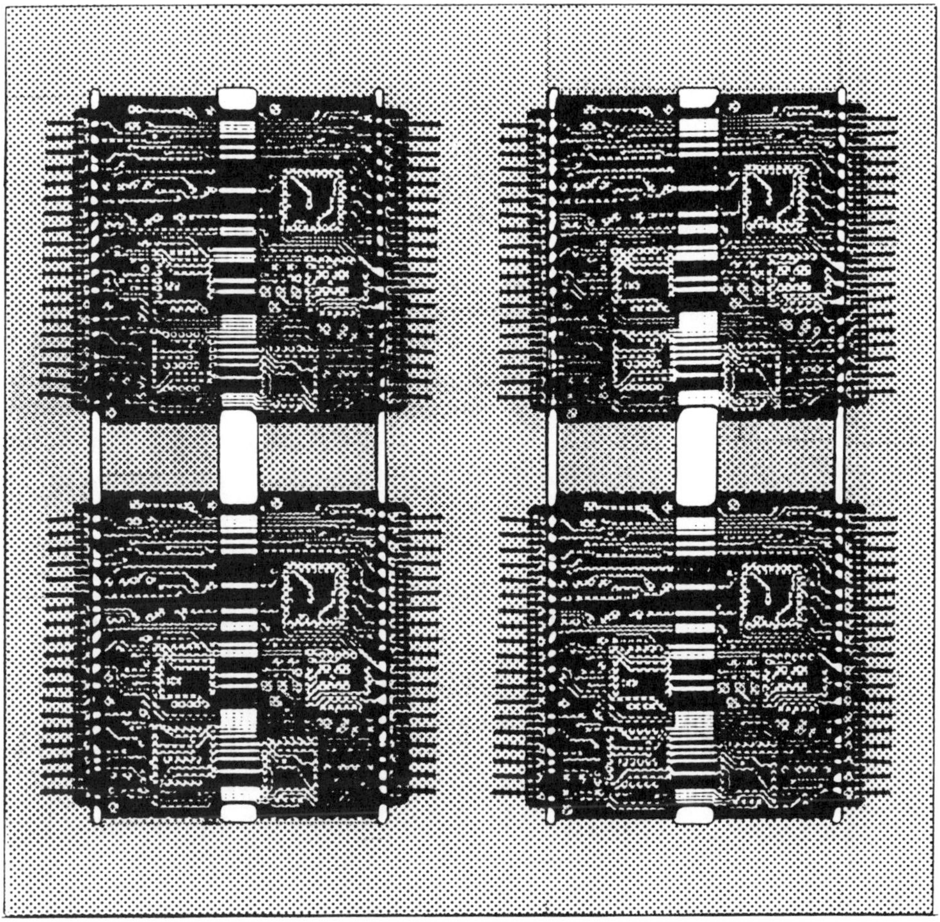

Fig. 6 Circuit patterns assembled with four flexible circuits. Copyright © Motorola, Inc.

This FR-4 rigid board carrier performs two functions.

First, the FR-4 glass epoxy substrate acts as a carrier to help process the flexible substrates during the attachment of SMDs and during wave soldering. Second, the FR-4 glass epoxy is selectively laminated to the top side of the flexible substrate. On this same plane several square windows are punched out of the rigid board material to facilitate the chip carrier attachment directly onto the flexible substrate surface.

The chip carrier geometrical patter is similar to the footprint used on a hybrid design layout, but with a unique difference. It has a hole at the end of each footprint pad. This hole is a plated-through-via (see Fig. 7).

On the far side of the flexible substrate, the holes have a minimum annular pad diameter. The reason for this special geometrical pattern is based on the theory that if a minimum pad area is encountered by the molten solder, there is only one way for that molten solder to migrate, and that is up the plated-through-via to the substrate's top side, as shown in Fig. 8.

The placement of the chip carrier on the flexible substrate is done with a nestle fixturing tool. This tool provides individual recesses to secure the chip carrier in place. It is recommended that a chip carrier package be selected for maximum accuracy when locating the chip carrier on the flexible substrate surface.

This fixturing tool can be keyed to the internal flex circuit tooling holes. Provisions for the individual set of guide pins in the fixturing tool will match the corresponding inter-

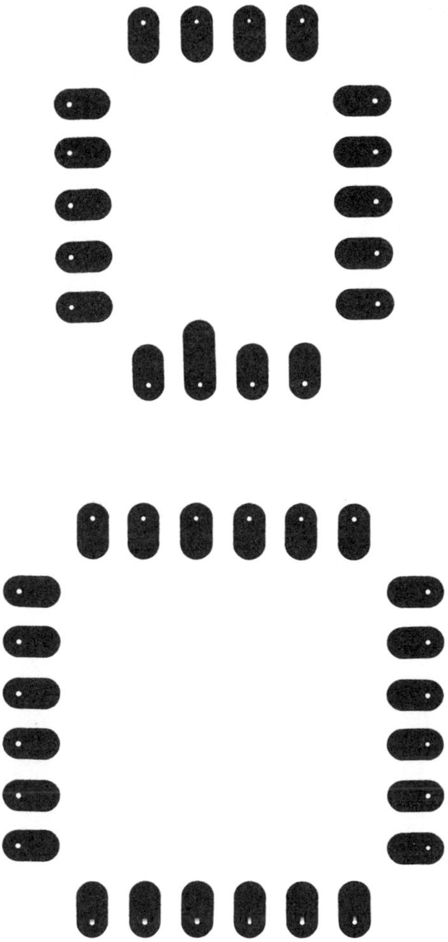

Fig. 7 Two examples of footprint pads with plated-through-vias.

nal tooling apertures found in the flexible substrates assembled on the rigid board corrier.

The flexible substrate tooling holes will also be used by the design engineer to line-up the automatic parts placement machine. This equipment will affix the passive SMDs

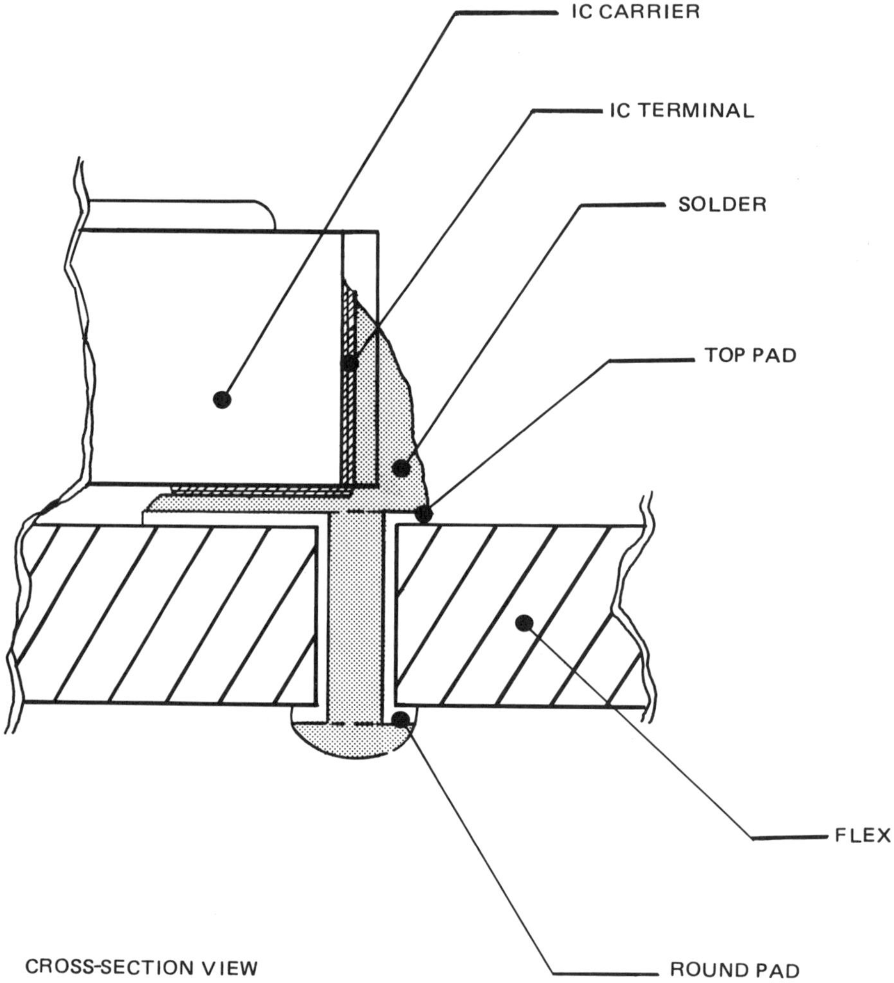

Fig. 8 Cross-section showing geometrical pattern which causes molten solder to migrate.

on the near side surface of the flexible substrate.

The SMDs are glued to the flexible substrate and then the parts will be cured, prior to the placement of the LCC and the other leaded components on the far side of the flexible substrate.

It is important to make sure that the flex circuit assembly is baked for a period of 30 min. at 150°F (65°C), prior to wave soldering to drive off moisture, because the Kapton polyimide film is hygroscopic.

After all the passive SMDs are affixed on the flexible substrate near side surface, and the remaining leaded components are secured on the far side surface, the assembly of the chip carrier on the nestle-fixture tool is the next item in the assembly process.

All chip carriers must be pretinned with solder and cleaned properly before they are placed in the nestle-fixture. Be sure that the chip carrier pin 1 orientation matches that of the flex circuit layout. The chip carrier should be placed top side down on the nestle-fixture crevices (see Fig. 9) for correct placement on the flexible circuit assembly.

Notice that a thin silicone cushion is used between the chip carrier's top side and the bottom side of the nestle-fixture crevice. This silicone cushion is needed to affix the chip carrier to the predetermined geometrical chip carrier footprint on the flexible circuit surface.

As the flexible substrate assembly is being compressed between the nestle-fixture and the wave solder fixture, the silicone cushion maintains the proper tension between the

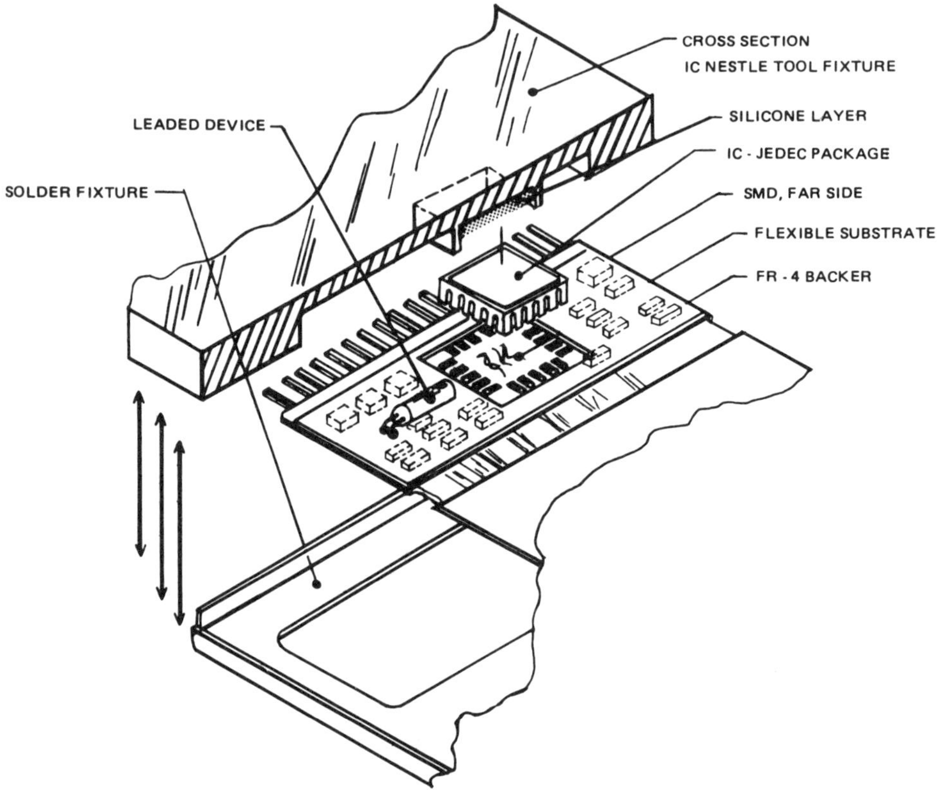

Fig. 9 Placement of the chip carrier on the flexible circuit assembly.

chip carrier package and the flexible substrate surface.

This tension is required because it gives the chip carrier package the ability to expand and contract as the flexible substrate assembly passes over the wave solder machine and the Kapton film begins to expand through the cutouts in the wave soldering fixture.

After the flexible substrate assembly is properly secured (between the two fixtures), the next step is to apply a sufficient amount of flux on the flexible circuit surface, making sure the through-vias used on the capillary soldering technique were properly wetted.

Ideally, dual-wave solder machine preheating should bring the temperature of the flexible substrate top side to between 200 and 220°F (93–105°C) just prior to entering the first solder wave. The flexible substrate assembly must be adequately preheated to eliminate flux generated gases and to limit the dwell time in the solder waves to no more than 3–4 sec. The dual solder wave conveyor speed is set to about 3 ft. per min. The wave height is set about ¼ in.

The first wave is a narrow, turbulent, jet-like fountain. The function of this wave is to insure that solder is forced into the chip array and reaches all the chip carriers attachment points via the plated-through-holes. Simultaneously, the pretinned solder begins to reflow onto the chip carrier's footprint on the flexible substrate surface (see Fig. 10).

The shape of the secondary wave is smooth and it flows in the same direction and at the same relative velocity as the flexible substrate assembly. This minimizes turbulence caused by chip components, and it provides the needed controllable exit conditions.

After the proper cooling time, the flexible substrate is cleaned and the leaded components are trimmed as needed.

It will be obvious that the chip carrier did not only adhere to the flexible circuit footprint pattern, but that it also was soldered on the chip carrier wall ridges, providing a good solder joint.

In conclusion, variations of the capillary

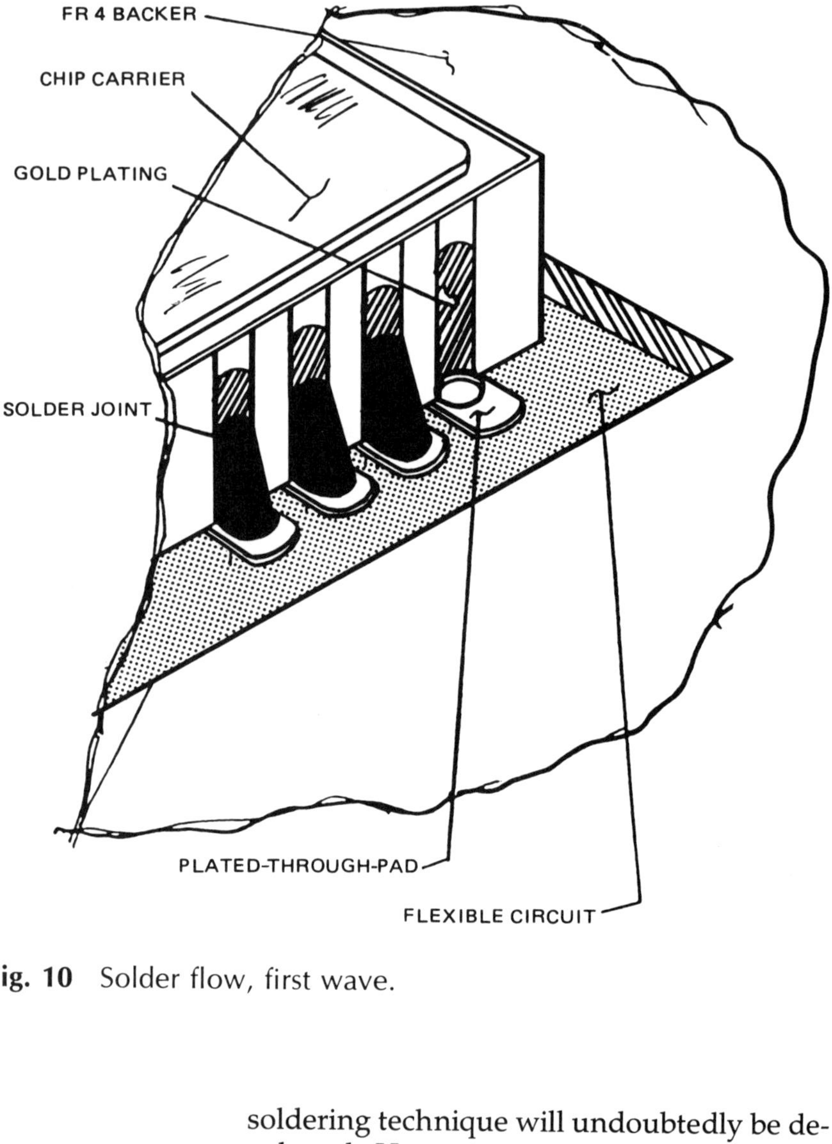

Fig. 10 Solder flow, first wave.

soldering technique will undoubtedly be developed. However, it is important that we understand that this procedure is only applicable when the proper flexible substrate thickness, the right fixturing tools as well as dual-wave soldering equipment are used.

5
WHAT LIES AHEAD?

As we look forward to what lies ahead, let's briefly review the material we have covered in this book.

We have seen how the best of two technologies, the hybrids method and the printed wiring board construction, have merged together and given birth to a new packaging technology—the flexible circuit.

Although we are just beginning to experience the impact of this new technology, the development of the surface mount technology on flexible circuit substrates of polyimide film offer the design engineer new design freedom, plus space and cost savings.

Until recently, mounting chip carriers directly onto a printed wiring board was the only way to take advantage of the high packaging density possible with these chip carriers.

Unfortunately, the thermal coefficient of expansion mismatch between the chip carrier and the conventional FR-4 glass epoxy printed circuit board material made such an ideal arrangement all but impossible and costly.

Using polyimide film, we can explore another technique to attach not only the chip carrier but also the chip itself to the flexible substrate surface as an integral part of the circuit construction.

Presently, flexible circuit substrates have been designed for high component density layouts which include both the use of SMDs and leaded components attached to a flexible substrate assembly.

The use of chip carriers or similar devices to house the chip requires a large surface area on the flexible substrate surface, limiting size reduction of the flexible substrate and/or the ability to design in more SMDs on the same flexible circuit layout.

The purpose of this attaching technique is to design a flexible substrate layout which is capable of interfacing with a tape automated bonding (TAB) system.

TAB is a semiconductor assembly and packaging system. The most visible aspect of the TAB system is the tape carrier. Polyimide, the tape material has high temperature properties, high mechanical strength, and a coefficient of linear expansion similar to copper. The tape carrier provides important functions to the TAB production process. It carries chips after inner lead bonding, provides the interconnection leads, and allows test of chips on tape.

Inner lead bonding is a process which simultaneously bonds all the leads on a TAB tape pattern to the bumps on one chip. Gold is the most widely used metal for bumps because it can be rapidly hard-soldered to tin coated copper leads in air without flux.

The bonding operation involves the acurate positioning of the chip under the tape pattern so that the leads are centered over the bumps on the chip, the preheating of the chip, and the bonding.

The outer lead bonding operation transfers the chip with leads in place to the flexible substrate's selected geometrical chip footprint pads. Outer lead bonding is a two-step process, usually done sequentially on the same piece of equipment.

The first step in the outer lead bonding process is to excise the chip with leads in place from the TAB tape (Fig. 1). The chip with leads attached is transferred to the flexible substrate in a way that precisely maintains the original relationship of the the chip with respect to its tape sprocket holes. This feature makes it feasible to preposition the flexible substrate bonding site at the exact place the chip will arrive.

A secondary lead-forming function may be included as part of the process. The leads must be shaped to sweep down to the flexible substrate level without shorting to the electrically active edge of the chip (Fig. 2).

Thermode/vacuum removal of soldered outer lead bonding chips provides a neat and inexpensive repair capability for tape automated bonding flexible substrates assemblies. Another advantage of this flexible circuit technique is the ability to screen print resistors on

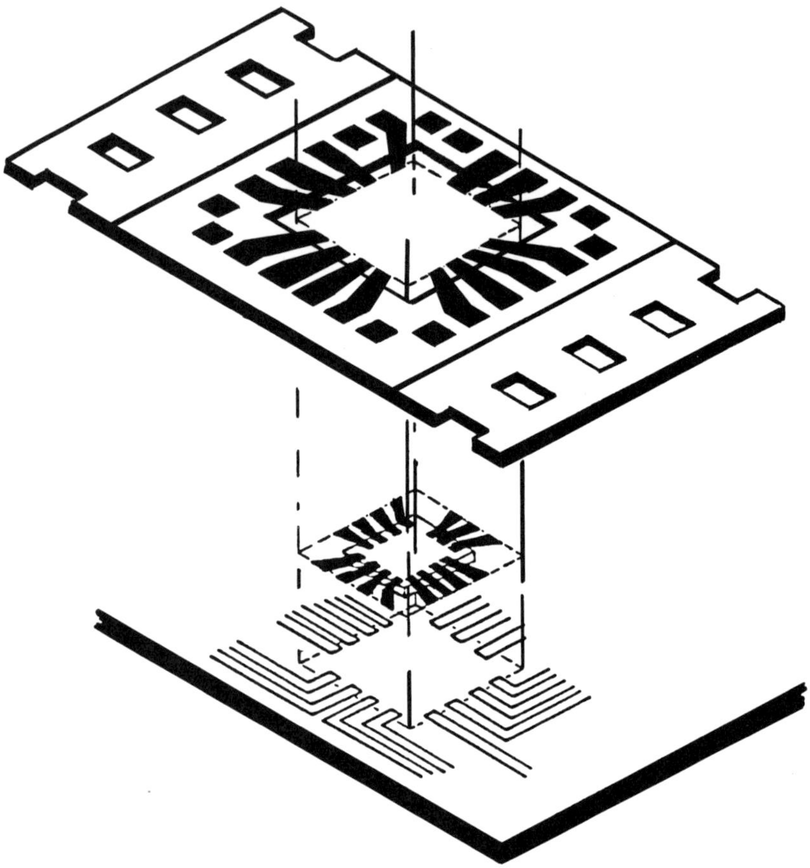

Fig. 1 Schematic representation of the excise and place function of outer lead bonding. The chip with its leads is transferred from the tape to the substrate. Courtesy of *Hybrid Circuit Technology*, November, 1984.

the flexible substrate surface. The only restriction at the present time is the resistor's tolerance, which is rated +/−20% of the base value. This restriction is mandated because no laser trim could be used on the polyimide film without affecting the integrity of the flex-

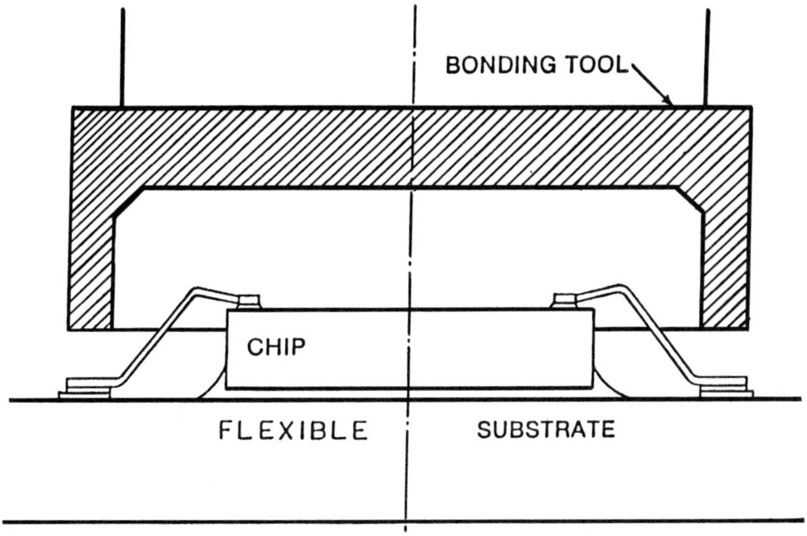

Fig. 2 Schematic representation of the outer lead bond. A hollow tool bonds all leads simultaneously. Courtesy of *Hybrid Circuit Technology*, November, 1984.

ible circuit assembly. However, any electronic circuit pull-up resistor can meet these design criteria and, therefore, more flexible circuit surface area will be available for other surface mount components.

Important new developments are coming from the raw material industry. New cover coat material construction techniques are being investigated, and new processes such as the new DuPont interconnection technique will allow the interfacing between two flexible substrate surfaces without the need for plated-through-vias. This technique will facilitate the construction of a double-sided, multilayer flexible circuit with better accuracy and more affordable unit price.

The possible new cover coat material versions, as well as better heat-controlled adhesives, will make the use of flexible circuits of polyimide film even more expedient and as an economical packaging technique of the not too distant future.

BIBLIOGRAPHY

Books

Gurley, S. Flexible Circuits Design and Applications. New York: Marcel Dekker, Inc., 1984.

Steinberg, David S. Vibration Analysis for Electronic Equipment. New York: John Wiley & Sons, 1973.

Periodicals

Ball, Thomas and Harry Capper, "Flexible Flat Circuitry Requires a Sound Interconnection System," *Electronic Packaging and Production*, March, 1984.

Brown, Don and Bruce Morgen. "Surface Mountable Components—An Update," *Electronic Packaging and Production*, January, 1984.

Capillo, Carmen. "Surface Mounted Assemblies Create New Cleaning Challenges," *Electronic Packaging and Production*, August, 1984.

Kraus, Joe. "High-Density Modules Suit Military Applications," *Design News*, October, 1983.

Lyman, Jerry. "Surface Mounting Alters the PC-Board Scene," *Electronics*, February, 1984.

Marks, Leonard. "CAD Speeds Flexible Circuit Design," *Electronic Packaging and Production*, December, 1983.

Markstein, Howard. "Laminates Support Technology Advances While Offering Alternative Choices," *Electronic Packaging and Production*, June, 1984.

______. "Pick-and-Place Swings Toward Surface Mount," *Electronic Packaging and Production*, January, 1984.

Murray, Bruce. "Choosing the Right Surface Mount Adhesive," *Machine Design*, October, 1984.

Ohr, Stephan. "Hybrid materials; New Inks and Substrates Offer New Problems . . . and New Possibilities," *Electronic Design*, August 17, 1979.

Peterson, Robert, Jr. "Components and Hardware," *EDN*, December, 1982.

Rahn, A. "Wave Soldering Leadless Components," *Circuit Manufacturing*, 1980.

Rima, Philip W. "The Basics of Tape Automated Bonding," *Hybrid Circuit Technology*, November, 1984.

Roth, Edward and Jeffrey Gerson. "A Comparative Analysis of Laminates Used in

'CC-4' Additive Substrative PC Bs,'' *Insulation/Circuits*, January, 1979.
Simpson, Stephen K. ''Flexible Circuit Symposium,'' DuPont Company, 1983.
Smookler, Lenny. ''New Additive Techniques Benefit Printed Wiring Boards,'' *Electronic Packaging and Production*, June, 1984.
Snyder, Ben H. ''Third Generation Flexible Printed Wiring,'' *Electronic Packaging and Production*, June, 1977.
Solder Manufacturers' Committee of Lead Industries Association, Inc., New York. ''Selecting the Proper Solder Alloy and Flux,'' *Electronic Packaging and Production*, July, 1979.
''Ultrapak SMSS for Surface Mounted Wave-Soldering,'' *Electrovert*, Bulletin 135, April, 1984.
Weiland, Donald. ''Hybrid Circuits Shed the 'Exotic' Label,'' *Machine Design*, June, 1978.
Wilkinson, Jeffrey. ''LCCC Implementation Procedures Enhance a Valuable Technology,'' *EDN*, June, 1982.

Standards and Specifications

IPC-D-249 Design Standard for Flexible One- and Two-Sided Printed Boards (Working Draft), 1984.
IPC-FC-243 Specification for Flexible Printed Wiring with Stiffeners, Single- and Double-Sided (Proposal).
U.S. Patent 4,472,238—Process using plasma for forming conductive through-holes through a dielectric layer.

U.S. Patent 4,489,999—Socket and Flexible PC board assembly and method for making.

U.S. Patent 4,517,050—Process for forming conductive through-holes through a dielectric layer.

U.S. Patent 4,533,188—Header and housing assembly for electronic circuit modules.

U.S. Patent 4,567,543—Double-sided flexible electronic circuit module.

GLOSSARY

A

- Active component A term used to describe electronic devices such as transistors and diodes.
- Adhesive dispenser The tool which is employed for affixing surface mounted devises with glue onto a substrate assembly.
- Anchoring tabs Extensions of the flexible circuit pad area which extend beneath the cover coat material to assist in holding down the pad to the base material.
- Annular ring A plated portion of conductive copper material which completely encircles a plated-through-hole.
- Artwork master An accurately scaled representation of a design layout which is used to produce the production master.

B

- Bad solder joint A solder connection exhibiting poor wetting and a grayish porous appearance due to either poor preheating, improper surface cleaning prior to soldering, or contaminated solder solution.

- Base material The insulating flexible material upon which the printed geometrical conductor pattern may be formed.
- Base material thickness The total thickness of insulating material excluding conductor, usually copper foil sheet.
- Board thickness The total thickness of the insulating base material, the metal conductive layer(s), and any optional plating or coating.
- BT material Bismaleimide triazine resin with a higher Tg spec.

C

- CAD Computer-aided design.
- CAM Computer-aided manufacturing.
- Capillary action The process in which a liquified substance can migrate from one plane to another via a connecting hole.
- Castellation The radial flatures around the edge of a ceramic chip carrier which are metallized to interconnect the conducting surfaces or planes within or on the top or bottom of the chip carrier. Connecting the solder joint of the chip carrier to the substrate usually involves the formation of a solder fillet to the castellation metallization.
- Center-to-center spacing The nominal distance between the centers of adjacent holes on a given geometrical pattern.
- CEM-1 A composite type of laminate material which incorporates an epoxy resin system and a flame retardant additive.

- CEM-2 A nonflame retardant version of CEM-1.
- Clad A fabrication process by which a thin metal foil layer has been laminated to one or both sides of an insulating base material.
- Clearance hole An aperture in the conductive geometrical pattern larger than, but coaxial with, an aperture in the insulating base material.
- CMOS Complimentary metal oxide semiconductor: a term used to describe a type of integrated circuit.
- Component density The area populated by electronic devices on a given surface.
- Component hole An aperture in the insulating base material used for the attachment and electrical connection of component terminations.
- Component lead The terminal wire which extends from a component and which serves as both mechanical support and electrical connection.
- Component side The side opposite to the solder side on which most of the leaded components will be mounted; also known as the far side.
- Conductive foil A thin sheet of copper foil that covers one or two sides of an insulating base material.
- Conductor A single conductive path fabricated from conductive copper foil on an insulating base material.
- Conductor size The copper foil width and thickness of a given conductive pattern on an insulating base material.
- Conductor spacing The distance between two adjacent conductor edges of isolated conductive copper foil patterns.

- Cover coat aperture The hole opening which is provided for direct access to the flexible substrate surface.
- Cover coat material The outer layer(s) of insulating material laminated to the conductive copper foil pattern on the insulating base material.
- CRM-5 A composite type of laminate material which incorporates a polyester resin system and a flame retardant additive.
- CRM-6 A nonflame retardant version of CRM-5.
- Current-carrying capacity The maximum current which can be handled by a copper foil conductor, based on width, thickness, and ambient temperature.

D

- Datum line A defined point, line, or plane used in a fabrication drawing to locate a given geometrical pattern.
- Daughter board The term used to identify a printed circuit board which plugs into a main board or "mother board."
- Delamination A separation between the insulating base material and the conductor layer or the cover coat layer.
- Device Another term used to identify a component or part.
- Dimensional stability A measure of dimensional change created by temperature, humidity, tensile stress, and age.
- Double-sided substrate A printed circuit substrate with a conductive geometrical pattern on both sides.
- Dual-wave soldering A wave soldering sys-

tem with two individual molten solder pots whose performance can be controlled independently.

- Dual in-line package A semiconductor device which terminates in two straight and equal rows of pins.

E

- Edge-flexible circuit contact A series of copper foil etched contacts on or near the narrow edge of a flex circuit substrate and intended for mating with a connector.
- Edge-to-conductor The minimum space required between the edge of a flexible circuit and the conductor edge.
- EMI shield Electro Magnetic Interference.
- Etch factor The ratio of the depth of an etch conductor's thickness to the amount of the lateral or undercut etch.

F

- Fabrication drawing A document that shows the dimensional limits or grid locations applicable to any or all parts of a printed circuit substrate, including the arrangement of conductive geometrical patterns, seize, type, and location of holes, as well as any other information necessary to manufacture a substrate.
- Flatpak package A semiconductor device which terminates in flat rectangular leads protruding from two sides of the device.
- Flex-O-Pak A trade mark name used by Motorola, Inc. to describe a proprietary flexible circuit package design.

- Flexible substrate A plurality of printed conductors, and leaded and leadless components utilizing flexible base and cover coat materials.
- Foot print (land pattern) A combination of lands intended for mounting a particular SMD. The metal pads on the substrate surface are arranged in the same pattern as the SMD leads or pads.
- Front-end charge A tooling cost incurred when setting up a hybrid production run.
- FR-4 A woven glass-cloth laminated material impregnated and bonded with an epoxy resin which incorporates a flame retardant additive.

G

- Geometrical pattern The name used to describe an arrangement of pads and/or conductors which represents the interconnecting pins of a given semiconductor.
- Glass transition temperature (Tg) The temperature at which an amorphous polymer changes from a hard and relatively brittle condition to a viscous or rubbery condition.
- Grid An orthogonal network of two sets of parallel equidistant lines used for locating points on a design layout.

H

- Hole breakout A condition in which an aperture is not completely surrounded by the land due to misregistration or poor artwork master tolerance.
- Hole density The quantity of holes per square

inch in a given printed circuit substrate.

- Hole pattern The geometrical arrangement of all holes in a printed circuit substrate.
- Hygroscopic Being capable of absorbing moisture.

I

- IC Integrated Circuit: This device can be found packaged in an LCCC, a plastic dual in-line, and a ceramic flatpak.

J

- JEDEC Joint Electronic Devices Engineering Council.

L

- Laminate A product made by bonding together two or more layers of insulating material and/or copper foil layer.
- Laminate thickness Total measured thickness of the metal-clad base material, single- or double-sided substrate, prior to any manufacturing process.
- Land area A selected area of conductive geometrical patterns usually used for the interconnecting or affixing (or both) of leaded and leadless components.
- LCC Leadless Chip Carrier: Made from any polyimide or FR-4 material which is nonhermetically sealable. Instead of leads, it has pads around its perimeter for connection to the substrate.

- LCCC Leadless Ceramic Chip Carrier: A hermetically sealable ceramic package in which a integrated circuit chip can be mounted to form a surface-mounted component. Instead of leads, it has pads around its perimeter for connection to the substrate.
- Legend, screen A format of letters, numbers, symbols, and patterns on the printed substrate primarily used to identify leaded and leadless components in assembly and replacement operations.
- LSI Large Scale Integrated Circuit.

M

- Mask, resist A coating material used to protect selected areas of a geometrical pattern from the effects of an etchant, molten solder, or plating bath.
- Metal-clad base material Base material covered with copper foil layer on one or both of the substrate sides.
- Minimum annular ring The minimum width of metal at the narrowest point between the edges of the hole and the outer edge of the land.
- Misregistration The lack of dimensional conformity of the position of a geometrical pattern with its intended position or with that of any other conductor layer of a substrate.
- Mother board The term used to identify a printed wiring board which accepts other boards that plug into its perimeter.

N

- Negative An artwork master in which the intended conductive geometrical patterns is transparent to light, and the areas which are to be free from conductive material are opaque.

- Nonwetting A condition whereby a substrate's surface has contacted molten solder, but the solder has not adhered to the entire surface. Base metal remains exposed.

O

- Opaque A material that renders a laminate sufficiently black so that the yarn or weave of the reinforcing material cannot be seen using either reflected or transmitted light.
- Outgassing Deaeration or other gaseous emission from a substrate assembly when exposed to reduced pressure, wave solder heat, or both.

P

- Packaging density Quantity of functions with leaded and leadless components per unit volume; usually expressed in qualitative terms, such as high, medium, or low.
- Pad A geometrical pattern which is designed to accept a leaded or leadless component termination.
- Passive component A term used to describe electronic devices such as resistors and capacitors.
- Photographic reduction dimension Dimension targets on the artwork master to determine the extent to which the artwork master is to be photo reduced to the 1 : 1 scale.
- Pick-and-place A system that takes compo-

nents from a feeding source, then orients and places them accurately on the surface of a substrate.

- Placement head That portion of the pick-and-place machine that interfaces directly with the component, using either vacuum, mechanical jaws, or both in transferring the component from the feeder to the surface of the substrate.
- Plated-through-holes An aperture in which electrical connection is made between two surface layers via the deposition of metal on the wall of the hole.
- Plating up The process which consists of the electro-chemical deposition of a conductive material on the base material after the base material geometrical pattern has been made conductive.
- PMOS (P+ Metal Oxide Semiconductor) A term used to describe an intergrated circuit.
- Positive An artwork master in which the intended conductive geometrical pattern is opaque to light, and the areas which are intended to be free from conductive material are transparent.
- Printed circuit layout A scale sketch that depicts the printed interconnect traces of a substrate, and the physical size and location of electronic leaded and leadless devices in sufficient detail to allow the preparation of a fabrication drawing and artwork master.
- Printed circuit substrate A conductive geometrical pattern comprised of a plurality of printed traces, printed components, or a combination thereof, all formed in and intended to be attached to a common base.

- Production art master A 1 : 1 scale art master which is used to produce one or more printed substrates within the accuracy specified in the artwork master.

R

- Reflow soldering A process in which a solder paste is deposited on pads of a printed circuit substrate and component leads are placed in the paste. When the solder has been heated to an appropriate temperature, it melts and forms a solder fillet which connects the pads and the component.
- Registration The degree of conformity of the position of a geometrical pattern with its intended position.
- Reworking The act of repeating one or more manufacturing operations for the purpose of improving the yield of acceptable parts.
- Right-angle connector A connector which terminates conductors at the edge of a printed circuit substrate, and brings the terminations out at right angles to the plane of the substrate.

S

- Schematic diagram A document which shows, by means of a block diagram or graphic symbols, the electrical connections among components and functions of a specific circuit arrangement.
- Signal An electrical impulse of a predeter-

mined voltage, current, polarity, and pulse width.

- Single-sided substrate A printed circuit substrate with a conductive geometrical pattern on one side only.
- SMD (Surface Mounted Device). Any active or passive device designated to be physically mounted to the surface of a substrate instead of by inserting leads into holes in the substrate.
- SMT (Surface Mount Technology). The entire technology of surface mounting, including components, substrates, pick-and-place systems, component attachment, soldering, and testing. SMT is a term generally applied when speaking of a substrate assembly which involves SMDs.
- Solder mask See mask, resist.
- Solder side The side of a printed circuit substrate which is opposite the component side.
- Soldering A process of joining metallic surfaces with molten solder without melting of the insulating base material.
- SOIC (Small Outline Integrated Circuit) A package in which an integrated circuit chip can be mounted to a surface-mount component. it is made from plastic material which can withstand high temperatures and which has leads formed in a gull-wing shape along its two sides for connection to the substrate.
- SOT (Small Outline Transistor) A plastic leaded surface-mount component in which diodes and transistors are packaged.
- Substrate Generic term for the medium used in interconnecting and supporting a group of electronic components; commonly ceramic or printed wiring board laminate material.

- Substrate, flexible Same as previously stated but made from a flexible material such as polyimide or polyester film.

W

- Wave soldering A process wherein printed circuit substrates are brought in contact with the surface of continuously flowing and circulating molten solder.
- Wetting The formation of a relatively uniform smooth, unbroken, and adherent film of solder to insulating base material.
- Wire bonding The process in which surfaces held together under moderately low static pressure are joined through the application of high frequency vibrations.

INDEX